CSR, Sustainability, Ethics & Governance

Series editors

Samuel O. Idowu, London Metropolitan University, London, United Kingdom
René Schmidpeter, Cologne Business School, Germany

More information about this series at http://www.springer.com/series/11565

Samuel O. Idowu • Catalina Sitnikov •
Lars Moratis
Editors

ISO 26000 - A Standardized View on Corporate Social Responsibility

Practices, Cases and Controversies

Springer

Editors
Samuel O. Idowu
Guildhall Faculty of Business and Law
London Metropolitan University
London, United Kingdom

Catalina Sitnikov
Economics & Business Administration
University of Craiova
Craiova, Romania

Lars Moratis
Antwerp Management School
NHTV Breda University of
Applied Science
Antwerp, Belgium

ISSN 2196-7075 ISSN 2196-7083 (electronic)
CSR, Sustainability, Ethics & Governance
ISBN 978-3-030-06481-5 ISBN 978-3-319-92651-3 (eBook)
https://doi.org/10.1007/978-3-319-92651-3

This Springer imprint is published by the registered company Springer Nature Switzerland AG
The registered company address is: Gewerbestrasse 11, 6330 Cham, Switzerland

Foreword

The introduction of neo-liberalism from the 1980s contributed to local and international shifts in economic practices and contracting nation-state welfare provision. These economic and social changes were unfolding at the same time as concerns about environmental change became increasingly well founded. The impacts on people and the planet became interwoven within rising anxieties and increasingly widespread political and consumer protests that targeted governments and institutions that were being held responsible for behaviour and policies that were viewed as being both short-term and irresponsible.

Corporate social responsibility, often building upon regional forms of philanthropy and religious sentiments, became a heavily utilized concept that was positioned to fill some of the gaps that deregulation and welfare cutbacks had caused. Businesses, especially those with wider reaches, and deeper financial pockets, such as transnational corporations, were no longer considered to just be part of the problem but also potentially part of the solution. In essence, corporate irresponsibility should not be inevitable. On the contrary, companies were now expected to extend their responsibilities from profit to include people and the planet. How companies responded to these greater demands upon their wider contributions has been the source of considerable contention. Practices and policies have varied markedly across the world and within and between primary, secondary and tertiary sectors. These variations partly reflect corporate motivations, the very different contexts in which businesses operate, and diverse stakeholder expectations, when these were taken into consideration. This lack of generic standards, guidance and restricted stakeholder contributed to the formation of ISO 26000.

At the onset ISO 26000 sought to bring a different approach to CSR; in the process the International Organization for Standardization (ISO) introduced a less rigorous standard setting system that they had hitherto implemented. Moreover, ISO 26000 was also more inclusive, the outcome of wide-ranging stakeholder engagement that helped avoid the more pointed criticism about the narrow development of other ISO standards. The outcome has been constructive tools for social responsibility guidance, providing instructive perspectives and principles that can be integral to sustainable CSR programmes. Nevertheless, the lack of certification that underpins other ISO standards has meant that in some respects the same concerns about validity, credibility and consistency that preceded ISO 26000 have continued in the years after its introduction. In other words, questions continue to be asked by CSR advocates and cynics alike about the intentions and outcomes of CSR. Following this reasoning leads to the question about the extent to which businesses can be held accountable for outcomes that do not fit neatly into a profit and loss account.

Despite this uncertainty and the ongoing challenges to CSR practices, there has been very little academic and practitioner analysis of the interplay between ISO 26000 and CSR. Therefore Idowu, Sitnikov and Moratis' *ISO 26000: A standardized view of Corporate Social Responsibility: Practices, cases and facts* is very timely, helping to enhance our understanding of the impacts and challenges of this CSR standard across a range of geographical settings; indeed, the range of regions and topics selected is in accordance with the stakeholder engagement that was so instrumental in the original formation of ISO 26000. The findings of this edited collaboration highlight the diversity of implementation, the ways in which local cultures and nature of businesses shape practices, problems that are encountered and some of the successful outcomes and benefits that ISO 26000 guidance is contributing to. Therefore, the book provides some incisive answers to questions swirling around ISO 26000 more specifically and CSR more generally while also making us better informed to help overcome difficulties that continue to restrict potentialities for sustainable, socially responsible development.

Head of School, School of Applied Social Studies, Robert Gordon University, Aberdeen, UK
Spring 2018

Stephen Vertigans

Preface

Organizations around the world and their stakeholders are becoming increasingly aware of the need for and benefits of socially responsible behaviour. The goal of social responsibility is to contribute sustainably to sustainable development.

On this basis, ISO 26000 is intended to support organizations on their road towards fulfilling sustainable development. The standard encourages organizations to overcome the threshold of a simple legal compliance, recognizing that compliance with law is a fundamental duty of any organization and an essential part of its social responsibility programmes. After 10 years of studies and negotiations, ISO 26000 was the result of an internationally innovative and unique multistakeholder process. It opened up the road to a new category of standards, based on participatory and consensual approach, aimed at drawing up recommendations in the form of "guidelines" and nonconformity requirements to a particular management system, such as the certifications proposed by other standards for example ISO 9001, ISO 14001, etc. The standard aims to promote a common understanding of social responsibility. When focusing on ISO 26000, organizations must take into account issues surrounding social, environmental, legal, cultural, political and organizational diversity, as well as differences in economic conditions, maintaining at the same time, being consistent with international norms of behaviour.

In the world of standards, ISO 26000 has proved lately to be an interesting and important approach. Although voluntary (with no third-party verification and certification), there are countries which have developed different versions of the standard to which organizations are voluntarily able to certify their SR processes. Seen and perceived as a voluntary standard that does not include any specific requirements, it was difficult to figure out the correct way to implement it. Nevertheless, taking into account the fact that even if resistance was manifested to building a management systems structure for ISO 26000, the future revisions and use will take place through incorporating it into other management systems standards such as ISO 9001 QMS and ISO 14001 EMS, which are already in place.

Therefore, this book attempts to provide readers with a detailed image of the ISO 26000 standard of SR, taking into account its focus on stakeholders and its seven main topics (organizational governance, human rights, labour conditions, the environment, business practices, consumer issues and social involvement). Addressing the 7 main topics in an integrated way, the book provides information on their focus and objective, their relationship with social responsibility, principles, considerations and related actions from ISO 26000 viewpoint.

The book underlines the key aspects and the most debatable issues of the standard with a focus on the fact that its core topics are correlated to the sustainable business practices. The cases, practical examples and results from organizations using the ISO 26000 guidelines are based on the fact that they can envision CSR in ways that fit their operational activities, stakeholders and the environment—and, by doing that, envision a better future.

London, UK
Craiova, Romania
Breda, The Netherlands/Antwerp, Belgium
April 2018

Samuel O. Idowu
Catalina Sitnikov
Lars Moratis

Acknowledgements

We are eternally grateful to a number of people who have helped us to realize the goal of publishing a book on ISO 26000. Our first thank you goes to all our contributors who have worked tirelessly to meet all the stringent requirements we had put in place to ensure that their papers passed our quality tests. We are also grateful to all our friends and colleagues who have supported this project.

We extend our sincere gratitude to our respective families for their support during the hard work which resulted in the book.

We would also like to thank our publishing team at Springer headed by the Executive Editor Christian Rauscher, Barbara Bethke and other members of the publishing team who have supported this project.

Finally, we apologize for any errors or omissions that may appear anywhere in this book; please be assured that no harm was intended to anybody. Causing harm or discomfort to others is simply not the spirit of corporate social responsibility.

Contents

List of Editors and Contributors

About the Editors

Samuel O. Idowu is a Senior Lecturer in Accounting and Corporate Social Responsibility at London Guildhall School of Business and Law, London Metropolitan University, UK. He researches in the fields of corporate social responsibility (CSR), corporate governance, business ethics and accounting and has published in both professional and academic journals since 1989. He is a freeman of the City of London and a Liveryman of the Worshipful Company of Chartered Secretaries and Administrators. Samuel is the Deputy CEO and First Vice President of the Global Corporate Governance Institute. He has led several edited books in CSR; he is the Editor-in-Chief of two Springer's reference books—the *Encyclopedia of Corporate Social Responsibility* and the *Dictionary of Corporate Social Responsibility*, an Editor-in-Chief of the *International Journal of Corporate Social Responsibility (IJCSR)*, the Editor-in-Chief of the *American Journal of Economics and Business Administration (AJEBA)* and an Associate Editor of the *International Journal of Responsible Management in Emerging Economies (IJRMEE)*. He is also a Series Editor for Springer's books on CSR, Sustainability, Ethics and Governance. One of his edited books won the most Outstanding Business Reference book Award of the American Library Association (ALA) in 2016, and another was ranked 18th in the 2010 Top 40 Sustainability Books by *Cambridge University Sustainability Leadership Programme*. Samuel is a member of the Committee of the Corporate Governance Special Interest Group of the British Academy of Management (BAM). He is on the editorial boards of the *International Journal of Business Administration*, Canada, and *Amfiteatru Economic Journal*, Romania. Samuel has delivered a number of Keynote Speeches at national and international conferences and workshops on CSR and has on two occasions 2008 and 2014 won Emerald's Highly Commended Literati Network Awards for Excellence. To date, Samuel has edited several books in the field of CSR, sustainability and governance and has written seven forewords to CSR books. Samuel has served as an external examiner to the following UK Universities—Sunderland, Ulster, Anglia Ruskin, Plymouth,

Robert Gordon University, Aberdeen, Teesside University, Middlesbrough, Sheffield Hallam University and Leicester De Montfort University.

Catalina Sitnikov, PhD, is a Professor at University of Craiova in Romania, at the Faculty of Economics and Business Administration. She holds a doctorate degree in management since 2000, habilitation title in management since 2014 and since February 2015 been a PhD supervisor in management. She was for 3 years a Visiting Lecturer at Helsinki University of Technology, Lahti Center, Finland. Since 1995, she has been teaching undergraduate, master's and PhD students. She teaches quality management, total quality management and management. Her main research areas include management, strategic management and mostly quality management, instruments and models specific to the stages of quality planning, control and improvement, quality management strategies, ISO standards and CSR from the perspective of specific standards and instruments.

Lars Moratis, PhD, is Professor of Corporate Social Responsibility and Sustainable Business at Antwerp Management School and NHTV Breda University of Applied Sciences. He holds a PhD in management science and has worked in the field of CSR in both the academic and the business context. His research interests lie in the fields of sustainable business models, CSR strategy development and implementation, the credibility of corporate CSR claims, the adoption of CSR and sustainability standards, critical approaches towards CSR, the psychology of sustainability and responsible management education. He has authored and edited several books, chapters, academic and practitioner-oriented articles and opinion pieces on these topics.

List of Contributors

Selena Aureli Bologna University, Bologna, Italy

Mara Del Baldo University of Urbino Carlo Bo, Urbino, Italy

Cătălin Mihail Barbu Faculty of Economics and Business Administration, University of Craiova, Craiova, Romania

Rosane Aparecida Gomes Battistelle Department of Civil and Environmental Engineering, School of Engineering, São Paulo State University (UNESP), Bauru, SP, Brazil

Claudiu Bocean Faculty of Economics and Business Administration, University of Craiova, Craiova, Romania

Mark Anthony Camilleri University of Malta, Msida, Malta

Dan-Cristian Dabija Babeş-Bolyai University Cluj-Napoca, Cluj-Napoca, Romania

Rafael Mattos Deus Department of Production Engineering, School of Engineering, São Paulo State University (UNESP), Bauru, SP, Brazil

Caroline D. Ditlev-Simonsen BI Norwegian Business School, Oslo, Norway

Samuel O. Idowu London Guildhall School of Business and Law, London Metropolitan University, London, UK

Lars E. Isaksson Bond University, Gold Coast, Australia

Katja Lähtinen Department of Marketing, University of Vaasa, Vaasa, Finland

Monica Logofătu Faculty of Economics and Business Administration, University of Craiova, Craiova, Romania

Mirja Mikkilä Lappeenranta University of Technology, Lappeenranta, Finland

Nayan Mitra Kolkata, West Bengal, India

Lars Moratis Antwerp Management School, Antwerp, Belgium

NHTV Breda University of Applied Sciences, Breda, Netherlands

Ninel Nesheva-Kiosseva New University of Bulgaria, Sofia, Bulgaria

Radu Ogarca University of Craiova, Craiova, Romania

Camelia Olari Faculty of Economics and Business Administration, University of Craiova, Craiova, Romania

Diana Iulia Popa Babeş-Bolyai University Cluj-Napoca, Cluj-Napoca, Romania

Silvia Puiu University of Craiova, Craiova, Romania

Bruno Michel Roman Pais Seles Department of Production Engineering, School of Engineering, São Paulo State University (UNESP), Bauru, SP, Brazil

Catalina Sitnikov Faculty of Economics and Business Administration, University of Craiova, Craiova, Romania

Anne Toppinen Department of Forest Sciences, University of Helsinki, Helsinki, Finland

Karina Rabelo Ogasawara Vieira Department of Production Engineering, School of Engineering, São Paulo State University (UNESP), Bauru, SP, Brazil

Alice Tatang Widjaja Evelilia Company, Paris, France

ISO 26000—A Standardised View of Corporate Social Responsibility Practices, Cases and Facts: An Introduction

Samuel O. Idowu

Corporate social responsibility (CSR) has continued to reshape and transform a number of business practices, including the way corporate entities of the twenty-first century perceive and report on both their financial and non-financial activities to their stakeholders and the world at large. In the quest for responsible reporting of the non-financial aspects of organisations, the Global Reporting Initiative, a set of CSR reporting guidelines came into being in 1997, the Social Accountability 8000 (SA 8000) a social certification standard for factories and organisations across the globe came into being also in 1997, the UN Global Compact came unto the global corporate reporting scene in July 2000, there are a few others like them which are now available to global corporate entities when reporting on their non-financial activities. Social responsibility reporting has become important in recent times for accountability, transparency and good business practice reasons. Corporations of this era have understood the enormous benefits derivable when they provide to readers these social reports on how they are dealing with the environmental, social and governance (ESG) aspects of their operational activities Idowu and Papasolomou (2007). Many of these international reporting guidelines have been put in place to help organisations that aspire to be active and be perceived as being socially responsible in this regard in the quest to provide understandable social information to their stakeholders. Not only that these international guidelines have been put in place to facilitate easy comparability between two similar companies either in the same sector or those in different sectors entirely information contained in their social reports.

The ISO 26000 standard (which is the focus of this book) also contains reporting guidelines issued by the International Standards Orgranisation in Geneva, Switzerland. This non-mandatory international standard on CSR came into being in

S. O. Idowu (✉)
London Guildhall School of Business and Law, London Metropolitan University, London, UK
e-mail: s.idowu@londonmet.ac.uk

S. O. Idowu et al. (eds.), *ISO 26000 - A Standardized View on Corporate Social Responsibility*, CSR, Sustainability, Ethics & Governance,
https://doi.org/10.1007/978-3-319-92651-3_1

November 2010 after a multi-year and stakeholder-inclusive process. It was developed to help organisations to assess and address effectively those social responsibilities which are important to their operational activities, the environment, their stakeholders and and the world at large. The standard provides its guidelines from five perspectives. These perspectives are noted under five headings which are:

- Understanding social responsibility
- Principles of social responsibilty
- Recognising social responsibility and engaging stakeholders
- Guidance on social responsibility core subjects
- Guidance on integrating social responsibility throughout an organistion

This non-certifiable standard under the *Principles of social responsibility,* one of the five perspectives noted above provides information on how companies should report on seven fundamental aspects of their activities (or principles for) them to be fully engaged in an effective system of social responsibility. These are identified below:

- Accountability
- Transparency
- Ethical behaviour
- Respect for stakeholder interests
- Respect for the the rule of law
- Respect for international norms of behaviour
- Respect for human rights

As at the time of putting this paper together in April 2018, the website of the ISO 26000 notes that there are about 70,000 corporate members of the ISO 26000 based in 114 countries worldwide. This would appear to be a significant success made by the standard nearly 8 years after its coming into existence, the number of members can only continue to rise as the idea of reporting on CSR continues to permeate the length and breadth of planet earth. The website of ISO 26000 suggests that an organization's performance on social responsibility can influence, among other things:

- Competitive advantage
- Reputation
- The ability to attract and retain workers or members, customers, clients and users
- The maintenance of employee morale, commitment and productivity
- The perception of investors, owners, donors, sponsors and the financial community
- Relationships with companies, governments, the media, suppliers, peers, customers and the community in which it operates

A number of scholars Wood (1991), Idowu (2009) and some international organisations have argued that business and society do not operate in a vacuum—the two are interwoven and cannot operate effectively without each other. This fact is echoed on the Homepage of the *International Organisation for Standardisation* and is central to the ISO 26000 standard. Needless to say, business cannot operate without society because the stock of their employees will be drawn from members of society, their customers will be societal members, their lenders and suppliers will also exist in the society. Conversely, society cannot function without business as the goods and services societal members need in order to exist will come from organisations, the livelihood of of the working societal members will be derived from the wages and salaries they get from their employer organisations. It is therefore important that the two—*business* and *society* must co-exist in order to effectively meet the needs of each other.

Idowu and Aluchna (2017) argue that CSR has dynamically reoriented the global corporate scene, the global economy, the way corporate entities are managed and run by corporate executives, the realisation that all classes of stakeholder are important to corporate survival, the way we treat the environment and process our wastes over the last thirty or so years cannot be denied. The bar of responsibility has continued to be raised day in day out, simply because as we continue to innovate, new things come to our consciousness every time. Several factors have been responsible for this dynamic transformation which the world has experienced and continue to experience in this regard. It became apparent to us all that many of the financially and socially irresponsible actions which the old capitalist systems perhaps; unintentionally brought unto the corporate arena were unsustainable and damaging to humanity. It also became glaringly clear that many corporate and individual actions taken either consciously or subconsciously had high environmental, human, social and financial costs. This realisation has led to a number of advancements with regard to how corporate social responsibility is perceived and practiced in most economies of the world and how we conduct the day to day running of business. All these actions taken by corporations need to be reported and made known to all and sundry.

Interestingly, Cooper and Owen (2007) note that following from the fallout from Enron and similar affairs, reputation building appears to provide a primary motivating factor for companies going down the corporate social responsibility CSR path. This is why they argue that organisations such as the UK's Business in the Community (BitC) notes in their explanation of the *business case* for CSR that CSR could be used by companies to manage and influence attitudes and perceptions of their stakeholders, building their trust and enabling the benefits of positive relationshipsto delever business advantage. But is CSR or reporting on it about corporate image management? Scholars have argued that this is not the case. Moir (2001) notes CSR is really not about corporate image management or some other activity aimed predominantly at business benefits, it is in fact about socially and environmentally responsible behaviour. It is no longer about maximising corporate profit and making other stakeholders worse off. It is about doing the best for society and all its constituents regardless of whether they are animate or inanimate. Strategic CSR albeit is important to all corporate entities since it is about taking actions which are designed to benefit an entity and its stakeholders, but strategic CSR is only part of the components of CSR, it is really what has been described as the business case for CSR which was noted above in Cooper and Owen (2007) with reference to how the UK's Business in the Community perceives the business case for CSR. If corporate entities were to concentrate their CSR actions on strategic CSR, they will only be advancing their profit motive and neglecting other desirable aspects of CSR. That would still be a refined version of old capitalism which led us to many of the current difficulties of this generation, that's the view expressed in Idowu and Aluchna (2017).

These—and other relevant CSR issues—are part and parcel of the ISO 26000 standard. This book has ISO 26000 as its focus and aims to shed light on the discourse and practice of CSR from the perspective of ISO 26000 and reflect on the ISO 26000 standard itself. When one views ISO 26000 as an important point in the development of CSR (or, in more general terms, the relationships between business and society and the responsible behaviour of the former in the latter), it is worthwhile to investigate this standard from multiple angles, see how it is applied in practice, and how it may be shaping organizations' CSR practice.

Chapters in this book have been divided into three parts, each part containing papers which focus on similar themes to enable readers to decipher how different areas of ISO 26000 are forming. Part I—Standardised View of CSR Practices of ISO 26000—which is made up of five chapters, Part II—Standardised View of CSR Cases on ISO 26000—which is made up of five chapters and Part III—Standardised View of CSR Facts on ISO 26000—encompassed in two chapters.

The first of five chapters on Standardised View of CSR Practices of ISO 26000 entitled "ISO 26000 Implementation: The Case of Top Romanian Companies" by two Romanian scholars, Radu Ogarca and Silvia Puiu who explore the extent of ISO 26000 implementation in Romania and the means used by Romanian companies in this process. Their study examined the websites of the largest 25 companies in Romania in terms of turnover (at the level of 2015) and also in the CSR reports if they were public limited companies. The result of the study they note was quite surprising as only four of these companies make explicit references to the standard on their online channels addressed to the general public. Their failure to make reference to ISO 26000 in the CSR reports and on their websites does not implicitly mean that these companies are not implementing the standard, probably because they use other standards not ISO 26000. In another chapter in Part I Moratis and Widjaja in a chapter on "*The Adoption of ISO 26000 in Practice: Empirical Results from The Netherlands*" took a bold step to research into issues relating to the adoption of comprehensive CSR standards, Moratis and Widjaja note that this is an uncommon area of research especially in terms of conducting an empirical research. The two scholars delved into this field of study with an empirical study of Dutch CSR professional companies. The aim of the study they note was to provide insights which complement existing literature. In addition, Moratis and Widjaja argue that the study may also guide standard developers and policy makers in the field of corporate social responsibility. In another chapter of Part I from Romania, Bocean and Sitnikov another team of two Romanian scholars pose a question which sets out to explore whether ISO 26000 is a case of implementation or a simple reporting practice in Romania. Bocean and Sitnokov went on in the study for the chapter to elucidate those companies in Romania that apply CSR tools and standards including ISO 26000, the chapter went on to establish the manner in which these Romanian companies communicate how they implement these CSR tools and standards.

Moving swiftly on to Part II which is about CSR cases on ISO 26000, the five chapters in the section from scholars based in six countries—Finland, Italy, India, Australia, Malta and Bulgaria provide great additions to the book on CSR and ISO 26000 cases from countries in Europe and Asia in different sectors. The first chapter of the part from Finland looks at four companies headquartered in Finland in the forestry and energy sectors. The chapter notes that forest-based companies are strongly focused on environmental issues and organizational governance as key priorities for implementing their sustainability agendas, while for example consumer issues and human rights receive less attention. The energy companies have met less public pressure towards its operations compared to the forest industry, these scholars note. Del Daldo and Aureli two Italian reputable scholars in a chapter focusing on the case of an Italian cooperative, named Camst, which represents one of the largest catering service firms in Italy. Del Baldo and Aureli note that the company already follows the SA 8000 and accounts for its actions by publishing the company's Social Report. These two scholars note in their case study analysis of Camst that an experienced user of standards like Camst can benefit from the introduction of ISO 26000, whose adoption has led to an improvement in existing practices and a greater integration of social responsibility into the organisation. Moreover, the analysis indicates that ISO 26000 is quite a general standard that can be adopted in several contexts. Del Baldo and Aureli argue. Still in Part II but going from Europe to Asia—India to be precise, in a chapter co-authored by *Isaksson* and *Mitra* of Australia and India respectively on the case of Toyota in India. The chapter assesses and exemplifies how one well known multinational company, TOYOTA, operates its CSR efforts in India with respect to the ISO 26000 principles and requirements from the following perspectives organizational governance, human rights, labour practices, the environment, fair operating practices, consumer issues; and community involvement and development. This paper, provides valuable insights for other MNCs operating in India via a benchmark of 'how-to' comply with the Indian CSR legislation based on Toyota's recognized and award winning efforts.

In Part III, the final part of the book which consists of just two chapters on "*Standardised CSR Facts on ISO 26000*" explores some facts about the implementation of ISO 26000 by companies in Norway and Brazil. The first chapter of the section by Ditlev-Simonsen on "*Expectations Versus Applications—Five Years with ISO 26000 in Norway*" argues that the purpose of the chapter is about understanding the contributions and results of ISO 26000 in Norway. The chapter compares application and expectations of ISO 26000 in two Norwegian companies in 2011 and 2016. The findings of the study undertaken by Ditlev-Simonsen reveal that ISO 26000 is no longer applied in either of the two companies. The main reasons for these two companies stopping to use ISO 26000 Ditlev-Simonsen notes are because of the availability of other tools such as the Global Reporting Initiative (GRI) and UN Global Compact, which have received much more attention and developed over time. ISO 26000 has been static, and by now almost forgotten. Further those interviewed in the study claim that ISO 26000 was an important tool for developing a common and international understanding of the CSR concept. The final chapter of the book by four scholars from Brazil argues that ISO 26000 aims to assist organizations in contributing to sustainable development and employing international standards of behaviour. These Brazilians scholars argue that this type of standard is attractive, but there are some challenges that organizations need to face before adopting it. Considering these challenges, the chapter presents some characteristics of ISO 26000 and the main drivers and barriers to adopting CSR initiatives as well social standards. Deus et al. reviewed the literature to present important findings about CSR, social standards, and the drivers and barriers of CSR. The chapter presents in detail the main drivers toward CSR, for example, stakeholder pressure, globalization, reputation, competitive advantage, decrease in risks around the business, government laws, improving the relationship between employees and the organization, improving the relationship between co-workers, compatibility with other standards, and company size. Deus et al. note that as there are drivers, there are also barriers to ISO 26000 implementation. Interestingly, the chapter presents some of these barriers to the implementation of CSR systems and standards. Some of these barriers they identified are lack of knowledge or awareness of CSR, lack of knowledge about how to implement CSR with the organization's strategy, commercial barriers which could be national and international, lack of sensitivity to the theme, and financial resources. The barriers can be a challenge for organizational management. Organizations can use the tools presented by the standards and take advantage of the drivers, reinforcing their convictions and justifying their actions in order to implement CSR initiatives. These are important facts that organisation starting out in the journey to implement ISO 26000 can take congnisance of.

It is hoped that all the twelve chapters of the book will provide information to all our global readers about how the CSR standard of ISO 26000 has continued to improve the (reporting) practices on CSR since November 2010 when it was presented to the global community and how CSR has continued to reorient and shape the global business arena.

References

Cooper, S. M., & Owen, L. O. (2007). Corporate social reporting and stakeholder accountability: The missing link. *Accounting, Organisation and Society, 32*, 649–667.

Idowu, S. O. (2009). Corporate social responsibility from the perspective of corporate secretaries. In S. O. Idowu & W. Leal Filho (Eds.), *Professionals' perspectives of corporate social responsibility*. Heidelberg: Springer.

Idowu, S. O., & Aluchna, M. (2017). The dynamics of CSR: An introduction in the dynamics of corporate social responsibility: A critical approach to theory and practice. In M. Aluchna & S. O. Idowu (Eds.). *The dynamics of CSR*. Cham: Springer.

Idowu, S. O., & Papasolomou, I. (2007). Are the corporate social responsibility matters based on good intentions or false pretences? An empirical study of the motivations behind the issuing of CSR reports by UK companies. *Corporate Governance: The International Journal of Business in Society, 7*(2), 136–147.

Moir, L. (2001). What do we really mean by corporate social responsibility? *Corporate Governance: The International Journal of Business in Society, 1*(2), 16–22.

Wood, D. J. (1991). Corporate social performance revisited. *Academy of Management Review, 16* (4), 691–718.

Samuel O. Idowu, is a Senior Lecturer in Accounting and Corporate Social Responsibility at London Guildhall School of Business & Law, London Metropolitan University, UK. He researches in the fields of Corporate Social Responsibility (CSR), Corporate Governance, Business Ethics and Accounting and has published in both professional and academic journals since 1989. He is a freeman of the City of London and a Liveryman of the Worshipful Company of Chartered Secretaries and Administrators. Samuel is the Deputy CEO and First Vice President of the Global Corporate Governance Institute. He has led several edited books in CSR, he is the Editor-in-Chief of two Springer's reference books—the Encyclopedia of Corporate Social Responsibility and the Dictionary of Corporate Social Responsibility, an Editor-in-Chief of the International Journal of Corporate Social Responsibility (IJCSR), the Editor-in-Chief of the American Journal of Economics and Business Administration (AJEBA) and an Associate Editor of the International Journal of Responsible Management in Emerging Economies (IJRMEE). He is also a Series Editor for Springer's books on CSR, Sustainability, Ethics and Governance. One of his edited books won the most Outstanding Business Reference book Award of the American Library Association (ALA) in 2016 and another was ranked 18th in the 2010 Top 40 Sustainability Books by, *Cambridge University, Sustainability Leadership Programme.* Samuel is a member of the Committee of the Corporate Governance Special Interest Group of the British Academy of Management (BAM). He is on the Editorial Boards of the International Journal of Business Administration, Canada and Amfiteatru Economic Journal, Romania. Samuel has delivered a number of Keynote Speeches at national and international conferences and workshops on CSR and has on two occasions 2008 and 2014 won Emerald's Highly Commended Literati Network Awards for Excellence. To date, Samuel has edited several books in the field of CSR, Sustainability and Governance and has written seven forewords to CSR books. Samuel has served as an external examiner to the following UK Universities—Sunderland, Ulster, Anglia Ruskin, Plymouth, Robert Gordon University, Aberdeen, Teesside University, Middlesbrough, Sheffield Hallam University and Leicester De Montfort University.

Part I
Standardised View of CSR Practices on ISO 26000

ISO 26000 Implementation: The Case of Top Romanian Companies

Radu Ogarca and Silvia Puiu

1 Introduction

Many companies consider themselves as socially involved, with a strategy of social responsibility or at least with some periodical actions in this direction. Still, many times there is an important gap between the declarations and the actual initiatives. Implementing an international instrument regarding social responsibility could bring more credibility for the company in relation to its stakeholders and at the same time it could help it to introduce some principles and patterns of social responsibility management, respectively to measure the activities in this area using specific performance indicators. Such an instrument is ISO 26000 standard.

According to ISO 26000 Post Publication Organisation—PPO (2016), ISO 26000 is an international standard created for offering guidelines to all entities, no matter their size or core activity, regarding the social responsibility and their impact in the community. It is very important to understand that the standard offers guidelines, suggestions, recommendations and not compulsory or mandatory requirements. Another aspect is that ISO 26000 is not certifiable and organizations should avoid using terms like: "certified", "verified", "compliance" or "requirements" (ISO 26000 PPO SAG 2012).

Seferian (2013) also highlights that a communication from ISO 26000 PPO SAG regarding the misuse of some terms related to this standard is not usual, this meaning that some consultants were offering certificates and companies were interested to buy them. Moratis and Cochius (2011) suggest that there is a need for companies to want to get a certificate for this standard. So, there appears the question if these practices should be reported or a revision of the standard should include a certifiable standard if the market wants that. The text of ISO 26000 standard which states that

R. Ogarca · S. Puiu (✉)
University of Craiova, Craiova, Romania

S. O. Idowu et al. (eds.), *ISO 26000 - A Standardized View on Corporate Social Responsibility*, CSR, Sustainability, Ethics & Governance,
https://doi.org/10.1007/978-3-319-92651-3_2

this is not certifiable is clear, but should it remain unchanged? The history of the standard is not so long and revisions were scheduled to happen every 3–4 years. At the last review (2013/2014), it was established that the standard will remain unchanged until the next revision in 2017.

The standard comprises seven core subjects relevant for all organizations: organizational governance, human rights, labor practices, environment, fair operating practices, consumer issues, community involvement and development. Its main principles are: accountability, transparency, ethical behavior, respect for stakeholders, law, international norms and human rights.

ISO 26000 PPO offers periodical training materials in order to support all those who are interested in applying this standard. The recommendation is to start from the top (with the company's mission and vision), integrate social responsibility (SR) into procedures, establish the priorities (after a gap analysis between the desired situation for the seven principles and the current one), the responsibilities in the sphere of influence, perform due diligence (clause 7.3.1) and communicate SR to the stakeholders.

Bernhart and Maher (2011) consider that it is useful to build on existing management systems or other standards already certified, because ISO 26000 comprises guidelines that are also included in other standards like ISO 14001 or ISO 9001. It is very important that the standard highlights the fact that it can be used by every company that wants to improve its SR, no matter the size.

According to Sitnikov and Bocean (2012), ISO 26000 "is emblematic for the growing trend of regulation and intervention in business". Banarra (2014) published an article about certification as the best way to murder ISO 26000 standard. Many companies prefer to get a certifiable standard in order to prove they are socially responsible, but ISO 26000 is about a continuous development, about choosing the issues that are applicable to a company and the intent is beyond any certification.

Hahn (2013) mentions that a few national standard organizations created certifiable standards based on ISO 26000, but these are only available in that country, meanwhile ISO 26000 is worldwide known: the Brazilian NBR 16001, the Spanish RS 10, the Australian AS 8003, the Austrian ONR 192500, the Danish DS 49001.

Guertler (2013) appreciates that ISO 26000 should be used in accordance with the characteristics of each culture, company, country, region. The standard offers guidelines but not a specific way to use it. The author considers that ISO and the national standardization bodies (NSB) who were involved in the creation of this standard should show how they applied the standard in order to be followed as good practices by others. Another aspect highlighted by the author is that the standard is not generally accepted by companies, especially SMEs because the standard shows what to do to be considered socially responsible but doesn't show why you should use it.

Zinenco et al. (2015) argue that "in the absence of a public database" regarding the implementation of the standard, it is very difficult to know the exact number of the companies which adopted it. The authors mention that European countries predominate among the countries adopting this standard and also show an active use of other standards too. Their research tried to explain if the standard is redundant or complementary with the other standards issued by ISO or other bodies. ISO 26000 by itself cannot lead to fully socially responsible companies if it is not accompanied by other tools and instruments or suffers from a lack of promotion.

After the first revision between October 2013 and March 2014, ISO (2015) summarized a few recommendations to be taken into account at the next revision: make it shorter, improve guidance on organizational governance, risk management, prioritizing issues, health, economic issues, applying international norms. Some of them were anticipated by Guertler (2013), being a reflection of the fact that most entities are SMEs and it is difficult for them to see the full potential of the standard or to have the knowledge to apply it. Castka and Balzarova (2008) also made some predictions regarding the spread of ISO 26000: companies will adopt it if it is recognized by their stakeholders, multinationals will be first adopters and will require the same from their partners, and the adoption will be higher in countries supporting the standard.

There are currently no public data or any research by academics strictly on the issue of ISO 26000 implementation at companies' level in Romania. In this context, the aim of our research is to identify the degree and the ways in which companies in Romania apply ISO 26000 standard.

2 The Romanian Background for the Implementation of ISO 26000 Standard

Romania is one of the countries which adopted ISO 26000 and Asociatia de Standardizare din Romania (ASRO) is the Romanian NSB. It has a publishing house (Standardizarea) and a blog where it offers useful information regarding standards. The NSB issues monthly magazines available online, publishing articles in order to raise the awareness on standardization, SR and offer examples of good practice.

Soon after the first publication of ISO 26000, ASRO promoted the brochures of ISO regarding the standard (ISO 2010a, b) which were aiming to briefly explain what the standard is about, what benefits has, why companies should be interested in applying it and how it can be used.

According to CES Romania (2014), there was created a simplified version of the standard ISO 26000 for NGOs within the Societal Project. The project led to the creation of the network Societal in which there are 500 NGOs that apply the standard in all their projects. Societal also launched the brand "eveniment responsabil" offering recognition to the events that are socially responsible. Another network was Reteaua RSC—Actionam responsabil (318 members) which issued a digital magazine Social Responsibility Review where there were published articles on CSR and standardization, including ISO 26000, in order to raise the awareness on these problems.

It is very difficult to have a thorough picture on ISO 26000 implementation in Romanian companies and in general on their CSR activity, because till recently there did not exist a legal obligation regarding nonfinancial performance reporting. Under these circumstances, according to the local directory of sustainability reports (published on csrreport.ro/director-raportori), in Romania there are only 22 companies (all from the private sector) that published at least a sustainability report between 2001 and 2015. Starting from January 2017, according to the Minister of Public Finance Order no. 1938/2016, which transposes in the Romanian legislation the provisions of EU Directive 95/2014, all public entities which have more than an average of 500 employees during the fiscal year will also include in their management report a nonfinancial statement comprising of information regarding the environment, the community and the personnel, human rights, fighting corruption and bribery.

In the next paragraphs we will review a few studies that analyzed CSR activities developed by Romanian companies and also their reporting.

KPMG (2013) conducted a research on the top 100 companies by their turnover and the result for Romania was that 69% of them were reporting about CSR in 2013, an increase compared with 54% in 2011. Romania has an average reporting, meanwhile countries like Germany (67%), Belgium (68%), and Switzerland (67%) have a lower percentage.

KPMG (2015) conducted another study and Romania decreased a little in 2015, but still very close to 70%. The global average is 73%, because the companies with the highest turnovers have the potential to develop CSR and also the knowledge for reporting it. According to a research on Romanian CSR conducted by CSRmedia and EY Romania (2016), 76% of the respondent companies declared they have a CSR strategy. From these, 52% said they have a local strategy and 24% a strategy received from the parent company. 21% of the respondents said they do not have a strategy but still develop CSR projects. Asked about the tools they use for evaluating the impact of CSR on the community, 30% of the companies said they use their own evaluation systems, 20% use international standards, 16% use evaluations provided by NGOs and 27% of them do not measure the impact.

An analysis on 2013, 2014 and 2015 shows an important decrease from 53% to 30% in terms of using their own evaluations. More companies started to use international standards and reporting as a consequence of the events that are organized in Romania on this subject. Asked about the means used for communicating their CSR policies, most of the companies (66%) mentioned their website, 59% press releases (10% increase comparing with 2014), 59% social media (9% increase), 37% CSR platforms (13% increase), 36% yearly reports (7% increase) and 18% do not communicate in any way.

The Azores (2014) published a report on the top food retailers in Romania (by turnover) in terms of CSR. The retailer with the highest score (63.75%) was Auchan, because it develops projects on all three areas (social, environment and economy) and measures its activity. Kaufland and Mega Image were on the second position (35%), because they do not use measurable indicators like Auchan for presenting CSR performance. Kaufland, Lidl, Auchan and Mega Image have on their websites a CSR report and surprisingly also Elan-Trio and Annabella, national food retailers. The retailers that do not have such reports are Metro, Selgros, Carrefour, Rewe Group, Cora, Billa, Profi, Unicarm, Succes Nic.

The Azores (2015) published Romania CSR Index 2015 using the data from the top 100 companies by turnover in Romania. The report shows that only 8% of the companies (Auchan, Coca Cola, Ursus, Petrom, Heineken, Raiffaisen, Rompetrol, Telekom) published a CSR report on 2014 in Romanian, none of them being evaluated by an external auditor. The other companies, even if they do not publish a CSR report, present some aspects of CSR (governance, environment, human rights, employees, consumers, community, diversity, economic impact). Regarding the environment, 9% of the companies present their total energy consumption, but only Coca Cola and Heineken publish their objectives to reduce the energy consumption. Regarding the companies reporting about human rights, the average is 10% (Raiffaisen Bank and Petrom having the highest scores). Related to the employees, consumers and the relation with socially responsible suppliers, Coca Cola got the highest scores, training all its employees, developing the most sustainable product (Dorna water) 100% biodegradable and made of plants (30%) and including in all contracts with the suppliers an annex regarding the conditions their partners should respect in terms of human rights and working conditions. The global scores for the top 10 companies in Romania CSR Index 2015 were: Coca Cola HBC (87.14%), Raiffaisen (83.33%), Ursus (69.12%), Heineken (61.76%), Petrom (57.35%), Telekom (54.84%), Rompetrol (47.06%), ALRO (36%), CEZ Distributie (33.33%), Auchan Romania (33.33%).

Many of these aspects are included in the clauses of ISO 26000. So, it is possible that many companies already apply the standard (at least partially), but they do not know the content of it. At the same time, there are more international instruments and standards regarding social responsibility that could be classified as following (Szatmari et al. 2012): codes of conduct (United Nations Global Compact; OECD Principles of Corporate Governance; Tripartite declaration of principles concerning multinational enterprises and social policy—OIM; Declaration on fundamental principles and rights at work—OIM; Fairtrade); management standards (ISO 26000 and the AA1000 Series—for working environment and employees, market, community, environment; SA8000 and OHSAS 18001—for working environment and employees; ISO 14001—for environment; ISO 9001—for market); reporting instruments and standards (AA1000; Global Reporting Initiative). In these conditions, companies can use alternative solutions to ISO 26000 (this not even being certifiable).

Taking into account these general aspects, it is not surprising that there are very few studies and quantitative research on the implementation of ISO 26000 in Romania and that few companies are identified as applying the standard.

3 Analysis of ISO 26000 Implementation in Romanian Companies

In order to have a better picture on ISO 26000 implementation in Romania, we focused our attention on large companies. For determining the sample of our research, we used turnover as a selection criteria, which is characteristic both for the dimension of the firm and for its financial performance.

Though things are not so simple and clear, in the professional literature there is a strong opinion according to which large companies are more involved in CSR actions (Udayasankar 2008) and also that companies with financial performance are more prone to develop CSR projects (Margolis and Walsh 2001).

As a consequence, the research objectives are: identifying the first 25 companies in Romania by their turnover in 2015; make a research on their websites to see if they published something about ISO 26000; draw some conclusions and make some recommendations in order to increase the number of the companies implementing the standard. The research methodology is based on the documentation regarding the top 25 companies and their CSR reporting (including the information on ISO 26000), the analysis of the data gathered from their websites and the synthesis of the data to provide valuable results for researchers or other companies interested in this subject.

There is a difference between applying ISO 26000 and communicating about it, but we will assume that companies with such important turnovers know the difference and if they apply it, they probably communicate it to the stakeholders benefiting from it.

A similar study was conducted by Filip et al. (2012) which analyzed CSR management standards in 40 responsible companies. They found information regarding ISO 26000 only on the websites of Orange Romania and Transgaz, but ten of them had information regarding GRI.

In a similar research, Băleanu et al. (2014) analyze the web pages of the most profitable 50 companies in Romania in 2013, focusing on their transparency and non-financial reporting practices. As for ISO 26000 standard, this is being analyzed in the literature review part, but there is no mention of any company declaring to apply it.

We established three research questions (RQ), in order to draw some useful conclusions regarding the connection between the turnover and the capacity of the company to implement ISO 26000 and communicate it accordingly.

RQ1: Do the companies in top 25 have a dedicated menu on their website regarding CSR? If yes, is there something about ISO 26000?

RQ2: Is there any connection between the place they occupy in the top and the implementation of ISO 26000?

RQ3: What do companies implementing ISO 26000 have in common?

Finding an answer for these questions will help us better understand the current situation in Romania regarding the spread of the standard, identify the barriers and offer some recommendations for increasing the awareness on these issues. DoingBusiness.ro (2016) published a ranking of the companies in Romania by their turnover in 2015 and we can see the list with the first 25 companies in Table 1.

Automobile Dacia has on its website a specific menu for SR, focused on sustainable development, education, local community and environment. We can find out about some events, a policy for environment, but no annual reports on CSR or anything about ISO 26000. Renault Group has on its website a CSR report for 2015 but ISO 26000 is not mentioned there. In its previous reports, Renault Group (2011, 2012) states that all its CSR actions are in accordance with the guidelines of ISO 26000.

On the website of OMV Petrom (OMV Petrom Marketing and OMV Gas), we can find a specific menu for Sustainability. OMV Petrom (2012) mentions the report as being the first sustainability report realized in accordance with Global Reporting Initiative (GRI). Starting 2013, the company complies with UN Global Compact. Neither of these reports, nor the annual report mention anything about ISO 26000. We can also find on the website a social commitment policy where OMV Petrom presents its strategy focused on three major directions: education, environment and new types of energy and innovation. In the submenu list of Sustainability, we can see information about human rights, diversity, ethics and volunteering. Some of these aspects are part of ISO 26000 standard, but without mention.

Kaufland Romania has on its website a section for Responsibility with four subcategories regarding ecological measures, assortment, society and K Olympics for schools. There is no CSR report on the website and no references regarding ISO 26000. However, it is specified that customers are offered an increasing selection of organic products and labeled with Fairtrade logo. The company was remarked as one of the most important retailers in Romania by Capital (2016) and also received the award "The best sponsorship project of a large company" at Gala Oameni pentru Oameni, an event organized by Asociatia pentru Relatii Comunitare and The American Chamber of Commerce in Romania in 2015. Kaufland and OMV Petrom are in the top for the best CSR companies in Romania according to Capital (2016).

Rompetrol has an entire website related to its CSR projects—impreunapentrufiecare.ro. On the website of the group Rompetrol is part of—KMG International, we can find information about CSR and yearly reports (Economic Impact Assessment 2016; Annual Report 2014 mentioning CSR initiatives; 2013–2014 Sustainability Report based on GRI G4, none of them mentioning ISO 26000). According to Deloitte (2016), Rompetrol (KMG International) and Ursus won Deloitte Green Frog Award Romania (1st edition) and qualified for the regional competition, because they had outstanding sustainability reports.

British American Tobacco Romania has not a website so all the information was gathered from the website of the parent company which has a sustainability performance center and applies GRI guidelines. On the website, we can find yearly and sustainability reports. None of them was translated in Romanian or have any references to ISO 26000 (British American Tobacco 2015). British American Tobacco Romania was awarded the Top Employers Romania 2016 and Top

Table 1 Top 25 companies in Romania by the turnover in 2015

No.	Company	Turnover (lei)	No.	Company	Turnover (lei)	No.	Company	Turnover (lei)
1	Automobile Dacia	19,164,558,589	10	Orange Romania	4,585,755,248	19	OMV Petrom Gas	3,495,155,919
2	OMV Petrom Marketing	14,776,387,601	11	Metro Cash&Carry Romania	4,494,359,155	20	Mediplus Exim	3,360,058,016
3	Kaufland Romania	9,173,630,241	12	Auchan Romania	4,444,160,904	21	Vodafone Romania	3,330,641,989
4	Rompetrol	8,243,177,812	13	MOL Romania	4,421,926,935	22	Samsung Romania	3,308,303,515
5	British American Tobacco (Romania)	7,350,865,953	14	Dedeman	4,361,619,720	23	J.T. International (Romania)	3,219,410,302
6	Lukoil Romania	5,437,244,338	15	Petrotel Lukoil	4,341,890,040	24	Hidroelectrica	3,183,237,563
7	Carrefour Romania	5,149,498,472	16	Arcelormittal Galați	3,803,508,731	25	Autoliv Romania	3,051,791,239
8	E.ON Energie	4,839,538,137	17	Ford Romania	3,756,336,413			
9	Pludi Market	4,724,056,714	18	Mega Image	3,562,917,265			

Source: http://doingbusiness.ro/financiar/top/cifradeafaceri/

Employers Europe 2016 certification for the great conditions providing for its employees (Top Employers Institute 2016).

Lukoil Group Romania includes Lukoil Romania (commerce) and also Petrotel Lukoil (refinery). On the website of the company, there is a dedicated menu for CSR, where we can find information about the projects Lokoil implemented. Most of them are focused on environment, health, culture, education, community. The company also has a policy regarding quality, environment, health, occupational security (Lukoil Romania 2012). None of these data makes any references to ISO 26000 and there is no CSR report on the website. Lukoil (2014) published a Sustainability Report 2013–2014 and uses GRI G3 for reporting.

Carrefour Romania has a submenu on its website, regarding its main priorities on SR: community, employees, consumers, environment and suppliers. There are no CSR reports on the website and no references to ISO 26000. On the international website, there are more data regarding CSR, The Annual Financial Report for 2015 which includes also information about CSR and 2015 annual activity and responsible commitment report. Carrefour (2015) presents in its financial report that CSR of the company is based on ISO 26000. This is done according to the guidelines of the standard since 2012, according to Carrefour (2012). None of the reports is in Romanian.

E.ON Romania (Energie and Distributie) has a dedicated section on the website, where the company presents some directions related to corporate governance, market, environment, community, employees, but no CSR reports or references to ISO 26000. On the international website, we can find a dedicated menu for sustainability (strategy, environment, social, governance, integrity) and the sustainability report for 2015. According to E.ON (2015), since 2005, the company has prepared the sustainability reports complying with GRI guidelines. No references for ISO 26000 were found.

Pludi Market is part of Lidl Romania since 2010, so the activity of CSR is done in accordance with the principles of the company, at international level. On the website of Lidl Romania, we can find a submenu dedicated to SR, where we can read about the numerous projects developed by the company, but no CSR reports or references to ISO 26000. Lidl Romania was for the second time awarded at PR Daily's 2015 Corporate Social Responsibility Awards for the campaign "Doneaza pentru cei care salveaza", the only Romanian campaign awarded in the international competition (SMARK 2016).

On the website of Orange Romania, we can find a dedicated menu for SR, the report for 2014 in Romanian and a reference to the GRI G4 guidelines. No references for ISO 26000 were found. On the international website, we can find the CSR report for 2015 and also a note regarding the methodology used for reporting, in which Orange (2015) declares that the report takes into account GRI, ISO 26000, the OECD guidelines and the United Nations Global Compact (UNGC).

Metro Romania does not have a CSR menu, a CSR report or any references to ISO 26000 on its website. On Metro Group's website, there is a dedicated menu for Responsibility focused on some important directions for the company. According to Metro Group, its principles are based on UNGC's ten principles and the last CSR report from 2013 to 2014 is prepared in accordance with the guidelines of GRI G3. The 2014–2015 Annual Report includes data about sustainability but it is not focused on it.

Auchan Romania has on its website a dedicated menu for CSR, focused on its main priorities (employees, environment, business partners, and customers) and presents some projects it developed. According to Ardelean (2014), Auchan Romania published a concise CSR report for 2013–2014. The report states that the company respects the principles of UNGC. We found the report using the search engine, but if there is any on the website, it is not easy to find, the search box revealing zero results for the keywords related to this issue. On Auchan Holding's website, we can find more detailed and updated CSR reports, the last one is for 2015 and it respects UNGC's principles. No references for ISO 26000 were found.

MOL Romania has a dedicated menu for SR on its website presenting the main programs, but no references to ISO 26000. Still, the company publishes a yearly report "Implicare in comunitate" (the last one for 2014) where it presents its main SR activities. On MOL Group's website, we can find the annual reports, also including a section for Sustainability. According to MOL's Annual Report (2015), this is prepared in accordance with GRI G4 guidelines.

Dedeman is a company with 100% Romanian capital. The website presents some press releases from which we find out about some CSR projects but there is no dedicated menu for SR or any report related to it, or any references to ISO 26000. Certificates for ISO 9001 and ISO 14001 are presented on the website.

Arcelormittal Galati has on its website a menu for Community, where we can find out about the main CSR projects developed by the company, but no reports or references to ISO 26000. On the international website, we can find the 2014 CSR report for Arcelormittal Galati (in English), but nothing about ISO 26000. The reports are in accordance with UNGC's principles and GRI G4 guidelines.

Ford Romania does not provide many informations regarding CSR on its website, but in the submenu Sustainability we are given the link for the parent company, where CSR reports can be found. According to the last report of Ford (2016), CSR activity is reported in accordance with UNGC's principles and GRI G4 guidelines. No references to ISO 26000 were found on the Romanian or international website of Ford.

Mega Image has a dedicated menu for sustainability and some news about the Report on Sustainability, prepared by the parent company Delhaize. No references for ISO 26000 were found. Delhaize Group (2014) presents the Sustainability Progress Report 2014 for Mega Image and mentions that it was awarded 1st prize in the Community Support Category at Romanian CSR Awards 2014. Delhaize Group (2015) published the Sustainability Progress Report 2015 where it presents the situation in different countries, including Romania. The report mentions GRI reporting, but not ISO 26000.

Mediplus Exim has a submenu for Social Involvement where we can read about the company's projects and the vision related to this aspect, but no report or references to ISO 26000. Mediplus is part of A&D Pharma which has an entire division for SR—Sensiblu Foundation, where we can find out about the developed projects, but also no CSR reports or references to ISO 26000.

Vodafone Romania has on its website a menu for Social Involvement and also one for Vodafone Foundation which publishes yearly reports for the developed CSR programs (the last from 2014) but with no references to ISO 26000. On the international website, there is a menu for Sustainability where we can find the last Sustainability Business Report 2015–2016, where Vodafone (2016) states that it complies with GRI G4 guidelines. No references to ISO 26000 were found.

Samsung Romania has on its website a menu for corporate citizenship and also the annual reports in English (the last from 2014, also including a section for Global Citizenship) and the reports on sustainability in English, but with a summary in Romanian (the last from 2015). The reports are for the entire activity of Samsung Electronics and prepared in accordance with GRI G4 guidelines (Samsung 2015). No references to ISO 26000 were found.

JTI Romania does not have its own website so all information are gathered from the international website, where we can find a dedicated menu for CSR and also for Responsible procurement. We can find the Sustainability Report for 2015 (JTI 2015) in accordance with GRI G4 guidelines. Though, the reports of JTI (2012, 2013a, b) mention that since 2012, they were prepared having ISO 26000 as a framework.

Hidroelectrica has a submenu for SR, some annual reports (the last from 2011) and an environment report for 2011–2014. There are no reports for CSR activity or any references to ISO 26000 or GRI.

Autoliv Romania does not have its own website, but the international one provides information on CSR in a dedicated menu. The Annual Report for 2015 (Autoliv 2015) presents the most important aspects related to the company's CSR, but no references to ISO 26000.

4 Conclusions

From the 25 companies, only three have 100% Romanian capital (Mediplus Exim, Dedeman and Hidroelectrica), the others are multinationals or former Romanian companies that were bought by multinationals. Most of them have a dedicated menu on their website for CSR (under different names, like Sustainability, Community, Citizenship, Corporate Governance) with some exceptions (Metro Romania has a menu only on Metro Group's website, JTI and Autoliv have websites only for the parent company).

All these data are summarized in Table 2, where we can see that most companies that are under the umbrella of a parent company do not have CSR reports in Romanian or if there are any, these are not always updated.

Answering RQ1, we noticed that Dedeman is the only one which does not have a CSR menu, except some press releases regarding this aspect. Also, the Dedeman website includes a menu called "Dedeman in sport", where there are presented several projects financed by the company, all in sports, a great part of these being considered social responsibility activities.

Table 2 Analysis of ISO 26000 implementation in the top 25 companies in Romania

No.	Company	CSR analysis	No.	Company	CSR analysis	No.	Company	CSR analysis
1	Automobile Dacia	CSR menu, reports only on Renault group, ISO 26000	10	Orange Romania	CSR menu, reports on GRI G4, ISO 26000 (IW) OECD guidelines, UNGC	19	OMV Petrom Gas	CSR menu, reports on GRI G4, UNGC
2	OMV Petrom Marketing	CSR menu, reports on GRI G4, UNGC	11	Metro Cash&Carry Romania	CSR menu and reports only on IW, GRI G3, UNGC, Fairtrade (RIOBA brand)	20	Mediplus Exim	CSR menu, no reports
3	Kaufland Romania	CSR menu, no reports, except Fairtrade	12	Auchan Romania	CSR menu, reports on UNGC	21	Vodafone Romania	CSR menu and reports only on IW (GRI G4)
4	Rompetrol	CSR menu, reports only on KMG Int., GRI G4, UNGC	13	MOL Romania	CSR menu, reports on GRI G4 (IW)	22	Samsung Romania	CSR menu, reports on GRI G4
5	British American Tobacco (Romania)	Only an international website (IW), CSR menu, reports on GRI	14	Dedeman	Only press releases for CSR	23	J.T. International (Romania)	Only IW, CSR menu, reports on GRI G4, ISO 26000
6	Lukoil Romania	CSR menu, reports only on IW, GRI G3	15	Petrotel Lukoil	CSR menu, reports only on IW, G3	24	Hidroelectrica	CSR menu, environment reports
7	Carrefour Romania	CSR menu, reports only on IW, ISO 26000	16	Arcelor Mittal Galați	CSR menu, reports only on IW (GRI G4), UNGC	25	Autoliv Romania	Only IW, CSR menu, reports
8	E.ON Energie	CSR menu, reports on IW, GRI	17	Ford Romania	CSR menu, reports only on IW (GRI G4)			
9	Pludi Market	CSR menu on Lidl Romania	18	Mega Image	CSR menu			

Source: Own synthesis

The other companies have a CSR menu on the Romanian website or the international website. Most of them have CSR reports (more detailed and updated on IW) which mention the guidelines they use in reporting. 15 of the 25 companies mentioned they prepare their CSR reports in accordance with GRI G3 or G4. Renault Group (Automobile Dacia is part of it), Carrefour, Orange and JT International mentioned also ISO 26000 as a framework for their reporting.

Regarding GRI, we also analyzed the database of Global Reporting Initiative (http://database.globalreporting.org). We noticed that there are differences between the way companies consider complying with GRI standards and the way the international organisation assesses their reports: GRI G4—KMG International/Rompetrol (for the report in 2015), Petrom (the report from 2015); GRI G3.1—Petrom (the report from 2013); non GRI (there are no references that the sustainability reports are based on GRI Guidelines and GRI Standards)—Arcelor Mittal (the reports from 2014, 2015, 2016), Auchan Romania (the reports from 2013, 2014), Orange Romania (the reports from 2006, 2007), Petrom (the reports from 2012). We also noticed that some of the companies declaring on their websites that they comply with GRI standards do not have their reports in GRI database (Vodafone Romania, E.ON Energie, Samsung Romania).

For RQ2 and RQ3, there is no connection between the place they occupy in the top 25 companies in Romania and the implementation of ISO 26000 and what these companies have in common is that they are multinationals and use a general reporting at international level (three of them are French). So, the place on the Romanian top cannot influence these CSR decisions. The companies are multinationals, but the others which did not precisely declare that they use ISO 26000 are also multinationals, so no connection here.

Our results are in accordance with those in other similar studies conducted in European Union and also in Romania.

Thereby, a study of European Commission (EC 2013) conducted on 200 large companies (having more than 1000 employees) from 10 European countries (Romania not included), using the same methodology as in our research (analysis of websites, CSR reports and codes of conduct for each company) proved that 33% of the companies refer to at least one of the CSR standards. The most mentioned were UNGC (32% of the companies) and GRI (31%). ISO 26000 was mentioned by less than 10% of the companies. Most often, Dutch and French companies make references to this standard (as we noticed in our research too).

Ducu and Băndilă (2013) conducted a study on Romanian companies applying nonfinancial reporting standards (there are references to 22 standards including ISO 26000). 74% of the companies (out of the 72 which were analyzed) mention they apply at least one standard, the most mentioned being: GRI (50 companies), UNGC (7 companies), AA1000 (5 companies). ISO 26000 is not mentioned, but we consider this is somehow natural. In our opinion, ISO 26000 is focused on the implementation of SR, meanwhile, for example, GRI is focused on reporting performance indicators related to CSR.

There are numerous articles written by specialists and also by ISO (2014) and GRI (2011), where we can find comparative tables for the clauses included in ISO

26000 and GRI G4. As Wendt (2015) argues: “there is clear synergy between ISO 26000 and GRI, although ISO 26000’s focus is clearly broader than that of the GRI.”

The main limit of our study is that top 25 companies in terms of their turnover excluded from our analysis companies that are very active from a CSR perspective (Ursus, Coca Cola, Raiffaisen Bank). Under these circumstances, we intend to conduct a future research in a few years after the nonfinancial reporting will become mandatory in Romania for some companies, and to consider also other criteria for our sample, for example the number of employees.

References

Arcelormittal. (2015). *Arcelormittal Galati 2014 corporate responsibility report*. Retrieved from http://corporate.arcelormittal.com/~/media/Files/A/ArcelorMittal/sdr-2015/country-reports/Rom-2014-EN.pdf

Ardelean, A. (2014). *Raport de dezvoltare durabila Auchan Romania 2013*. Retrieved from www.csrreport.ro/rapoarte-csr-sustenabilitate/raport-de-dezvoltare-durabila-auchan-romania-2013.html

Auchan Holding. (2015). *2015 Activity and sustainable development report*. Retrieved from http://asp.zone-secure.net/v2/198/2213/64199/AUCHAN%2DRADD%2DGB%2D2015.pdf

Autoliv. (2015). *Annual report 2015 saving more lives*. Retrieved from www.autoliv.com/Investors/Financial%20Reports/AR2015.pdf

Băleanu, M., Irimie, S., & Irimie, S. (2014). *Corporate social responsibility in Romania: From conceptual frameworks to concrete actions*. Paper presented at the 4th review of management and economic engineering international management conference “The management between profit and social responsibility”, Cluj Napoca, Romania, 18–20 September.

Banarra. (2014). *Are we about to murder ISO 26000?* Retrieved from www.banarra.com/news/article/are-we-about-to-murder-iso-26000

Bernhart, M., & Maher, F. J. S. (2011). *ISO 26000 in practice*. Milwaukee: ASQ Quality Press.

British American Tobacco. (2015). *BAT annual report 2015*. Retrieved from www.bat.com/group/sites/uk__9d9kcy.nsf/vwPagesWebLive/DO9DCL3B/$FILE/medMDA87PVT.pdf?openelement

Capital. (2016). *Cei mai buni responsabili CSR*. Retrieved from www.capital.ro/cei-mai-buni-responsabili-csr.html

Carrefour. (2012). *2012 Registration document*. Annual financial report. Retrieved from www.carrefour.com/sites/default/files/Document%20de%20Reference%202012%20VENG.pdf

Carrefour. (2015). *2015 Annual financial report*. Retrieved from www.carrefour.com/sites/default/files/carrefour-registration_document_2015_annual_financial_report_0.pdf

Castka, P., & Balzarova, M. A. (2008). ISO 26000 and supply chains - on the diffusion of the social responsibility standard. *International Journal of Production Economics, 111*(2), 274–286.

CES Romania. (2014). *Rolul Consiliilor Economice si Sociale si al institutiilor similare din UE in promovarea conceptului de Responsabilitate Sociala Corporativa*. Retrieved from www.ces.ro/newlib/studii-ces/STUDIUL-Responsabilitate-Sociala-Companii.pdf

CSRmedia & EY Romania. (2016). *Tendinte si Realitati CSR in Romania 2016*. Retrieved from www.csrmedia.ro/studii-csr/

Delhaize Group. (2014). *Sustainability progress report 2014*. Retrieved from https://sustainabilityreport.delhaizegroup.com/2014/local-performance/romania-mega-image/

Delhaize Group. (2015). *Sustainability progress report 2015*. Retrieved from www.aholddelhaize.com/media/1944/sustainability-progress-report-2015-4.pdf

Deloitte. (2016). *Green frog award Romania*. Retrieved from www2.deloitte.com/ro/ro/pages/about-deloitte/articles/KMG-International-si-Ursus-Breweries-castigatoarele-Deloitte-Green-Frog-Award-in-Romania.html

DoingBusiness.ro. (2016). *Top companii dupa cifra de afaceri*. Retrieved from http://doingbusiness.ro/financiar/top/cifradeafaceri/

Ducu, C. & Băndilă, R. (2013). *Cercetare asupra standardelor internațional de raportare non-financiară – document de lucru al Coaliției pentru Afaceri Curate*. Retrieved from http://www.cleanbusiness.ro/?p=306

E.ON. (2015). *Sustainability report 2015*. Retrieved from www.eon.com/content/dam/eon-com/Nachhaltigkeit/CS-Report-2015/downloads/EON_Sustainability_Report_2015_final_PDF_EN.pdf

European Commission (EC). (2013). *An analysis of policy references made by large EU companies to internationally recognized CSR guidelines and principles*. Retrieved from http://ec.europa.eu/DocsRoom/documents/10372/attachments/1/translations/en/renditions/pdf

Filip, R., Iamandi, I. E., & Munteanu, S. M. (2012). Corporate social responsibility management standards in Romania: Application and evaluation at corporate level. *Business Excellence and Management, 2*(2), 35–46.

Ford. (2016). *Sustainability report 2015/16*. Retrieved from http://corporate.ford.com/microsites/sustainability-report-2015-16/doc/sr15.pdf

GRI. (2011). *GRI and ISO 26000: How to use the GRI guidelines in conjunction with ISO 26000*. Retrieved from www.globalreporting.org/resourcelibrary/How-To-Use-the-GRI-Guidelines-In-Conjunction-With-ISO26000.pdf

Guertler, G. (2013). *Is ISO 26000 ready for revision?* Retrieved from www.26k-estimation.com/Is_ISO26000_ready_for_revision-v6.pdf

Hahn, R. (2013). ISO 26000 and the standardization of strategic management processes for sustainability and corporate social responsibility. *Business Strategy and the Environment, 22* (7), 442–455.

ISO. (2010a). *ISO 26000 project overview*. Retrieved from www.iso.org/iso/iso_26000_project_overview.pdf

ISO. (2010b). *Discovering ISO 26000*. Retrieved from www.iso.org/iso/discovering_iso_26000.pdf

ISO. (2014). *GRI G4 guidelines and ISO 26000:2010. How to use GRI G4 guidelines and ISO 26000 in conjunction*. Retrieved from www.iso.org/iso/iso-gri-26000_2014-01-28.pdf

ISO. (2015). *The 2014 survey of post publication activities related to ISO 26000*. Retrieved from www.slideshare.net/StaffanS/iso-26000-ppo-meeting-with-advisory-group

ISO 26000 PPO. (2016). *ISO 26000 basic training material*. Retrieved from www.iso.org/iso/iso_26000_basic_training_material_annexslides.pptx

ISO 26000 PPO SAG. (2012). *ISO 26000 communication protocol*, N15, rev1. Retrieved from www.iso.org/iso/n15_rev_1_ppo_sag_-_communication_protocol.pdf

JTI. (2012). *Corporate social responsibility report*. Retrieved from www.jti.com/files/9613/5584/8273/2012_CSR_report_web_version.pdf.pdf

JTI. (2013a). *Corporate social responsibility JT group report 2013*. Retrieved from http://jti.com/files/4413/7173/6320/2013_CSR_report_web_version.pdf

JTI. (2013b). *Annual report 2013*. Retrieved from www.jti.com/files/1913/7457/3873/annual2013_E_all.pdf

JTI. (2015). *Sustainability report FY2015*. Retrieved from www.jti.com/files/6514/6469/4924/CSR_2015_INTERACTIVE_27052016_2.pdf

KMG International. (2014). *Annual report 2014*. Retrieved from www.rompetrol.com/sites/default/files/rap_an_en.pdf

KMG International. (2015). *2013–2014 Sustainability report*. www.rompetrol.com/sites/default/files/report_complete_simulated_en_redux.pdf

KMG International. (2016). *Economic impact assessment 2016*. Retrieved from www.rompetrol.com/sites/default/files/kmgi_study_en_preview.pdf

KPMG. (2013). *The KMPG survey of corporate responsibility reporting 2013*. Retrieved from www.kpmg.com/Global/en/IssuesAndInsights/ArticlesPublications/corporate-responsibility/Documents/corporate-responsibility-reporting-survey-2013-exec-summary.pdf

KPMG. (2015). *The KMPG survey of corporate responsibility reporting 2015*. Retrieved from www.kpmg.com/CN/en/IssuesAndInsights/ArticlesPublications/Documents/kpmg-survey-of-corporate-responsibility-reporting-2015-O-201511.pdf

Lukoil. (2014). *Sustainability report 2013–2014*. Retrieved from www.lukoil.com/materials/doc/reports/Social/Book_SO_eng.pdf

Lukoil Romania. (2012). *Politica si obiectivele in domeniul calitatii, mediului, sanatatii si securitatii ocupationale*. Retrieved from http://lukoil.ro/upload/pdf/Politica-si-obiective-Calitate-Mediu-SSM_2012.pdf

Margolis, J. D., & Walsh, J. P. (2001). *People and profits?: The search for a link between a company's social and financial performance*. Routledge: Psychology Press.

Metro Group. (2014). *Corporate responsibility report 2013–2014*. Retrieved from http://reports.metrogroup.de/2013-2014/corporate-responsibility-report/our-results/gri-index.html

MOL Group. (2015). *Annual report 2015*. http://annualreport2015.mol.hu/images/downloads/en/MOL-Group_Annual_Report_2015_EN.pdf

MOL Romania. (2014). *Implicare in comunitate 2014*. Retrieved from https://molromania.ro/images/mol_ro/pdf/Despre_MOL/Responsabilitate_sociala/Implicare_in_comunitate/Brosura_Implicare_in_Comunitate_MOL_Romania_2014.pdf

Moratis, L., & Cochius, T. (2011). *ISO 26000 the business guide to the new standard on social responsibility*. Routledge: Greenleaf Publishing.

OMV. (2012). *Raport privind dezvoltarea durabila pe anul 2012*. Retrieved from www.omvpetrom.com/SecurityServlet/secure?cid=1255753974868&lang=ro&swa_site=&swa_nav=&swa_pid=&swa_lang=

Orange. (2015). *Corporate social responsibility. Methodology note 2015*. Retrieved from www.orange.com/fr/content/download/37533/1150440/version/4/file/Note+methodologique+rapport+RSE+Orange+2015_Eng.pdf

Orange Romania. (2014). *Raport de responsabilitate sociala 2014*. Retrieved from www.orange.ro/about/2014_raport_csr.pdf

Renault Group. (2011). *An ambitious strategy, for a brand in full renewal.* 2011 Annual report. Retrieved from https://group.renault.com/wp-content/uploads/2014/07/renault_-_2011_annual_report.pdf

Renault Group. (2012). *A world of responsibilities*. 2012 CSR report. Retrieved from https://group.renault.com

Renault Group. (2015). *Global, local, committed.* 2015 CSR report. Retrieved from https://group.renault.com/wp-content/uploads/2016/07/rapport-rse-2015_en_.pdf

Samsung. (2015). *Sustainability report 2015*. Retrieved from www.samsung.com/common/aboutsamsung/download/companyreports/2015_Sustainability_Report.pdf

Seferian, N. (2013). *We're ISO 26000 certified. Say what?!* Retrieved from www.26k-estimation.com/ISO26000_Article_Responsible_Business_Issue_7.pdf

Sitnikov, C. S., & Bocean, C. G. (2012). Corporate social responsibility through the lens of ISO standards. *Business Excellence and Management, 2*(4), 56–66.

SMARK. (2016). *Lidl Romania - premiat pentru a doua oara in cadrul PR Daily's 2015 corporate social responsibility awards*. Retrieved from www.smark.ro/articol/34057/lidl-romania-premiat-pentru-a-doua-oara-in-cadrul-pr-daily-s-2015-corporate

Szatmari, I., Macovei, C. & Horia, C. (2012). Implementarea responsabilității sociale. *Ghid de instrumente şi tehnici*. Retrieved from http://undp.ro/libraries/projects/CSR/deliverables/Implementarea%20Responsabilitatii%20Sociale%20-%20Ghid%20de%20instrumente%20si%20tehnici.pdf

The Azores. (2014). *Top cei mai responsabili retaileri alimentari din Romania*. Retrieved from www.slideshare.net/responsabilitate_sociala/topul-celor-mai-responsabili-retaileri-alimentari-din-romania-in-2014?qid=4ceed540-b53e-42df-bb9b-efb79f913e87&v=qf1&b=&from_search=2

The Azores. (2015). *Romania CSR index 2015*. Retrieved from www.revistabiz.ro/res/wp-content/uploads/2016/01/Romania_CSR_Index_2015.pdf

Top Employers Institute. (2016). *Top employers Romania 2016*. Retrieved from http://ro.top-employers.com/en/companyprofiles/ro/british-american-tobacco-romania/

Udayasankar, K. (2008). Corporate social responsibility and firm size. *Journal of Business Ethics, 83*(2), 167–175.

Vodafone. (2016). *Sustainable business report 2015–16*. Retrieved from www.vodafone.com/content/dam/vodafone-images/sustainability/downloads/report2016.pdf

Wendt, K. (2015). *Responsible investment banking*. Cham: Springer.

Zinenco, A., Rovira, M. R., & Montiel, I. (2015). The fit of the social responsibility standard ISO 26000 within other CSR instruments. Redundant or complementary? *Sustainability Accounting, Management and Policy Journal, 6*(4), 498–526.

Websites

csrreport.ro/director-raportori
http://database.globalreporting.org
http://galati.arcelormittal.com
www.asro.ro
www.auchan.ro
www.carrefour.ro
www.daciagroup.com
www.dedeman.ro
www.eon-romania.ro
www.ford.ro
www.hidroelectrica.ro
www.kaufland.ro
www.lidl.ro
www.mediplus.ro
www.mega-image.ro
www.metro.ro
www.rompetrol.ro
www.samsung.com/ro/home/

Radu Ogarca, PhD is an Associate Professor at the Department of Management, Marketing and Business Administration within the Faculty of Economics and Business Administration at the University of Craiova, Romania. He has a Master's degree in business administration at University of Craiova and a PhD in Management from the Bucharest University of Economics Studies. Since 2002, he has been teaching Bachelor's and Master's students. He teaches Management, Small Business Management, Entrepreneurship and Business Ethics. His research is focused on business ethics, corporate social responsibility, entrepreneurship and social entrepreneurship. In December 2013, he graduated with a postdoctoral studies after a 36-month period in which he made a research on *small business ethics in Romania.* In 2014, he won a grant from the University of Craiova and made a research on *Ethics Management in Higher Education System of Romania* together with his co-author of his chapter in this ISO 26000 book—Silvia Puiu.

Silvia Puiu, PhD is a Lecturer at the Department of Management, Marketing and Business Administration within the Faculty of Economics and Business Administration, University of Craiova, Romania. She has a Master's degree in International Business Administration and a PhD in Management from the University of Craiova. She published her PhD thesis on Strategic Management of Retail Sector in Romania in 2012. In 2014, she conducted a research on *Ethics*

Management in Higher Education System of Romania together with PhD Associate Professor Radu Ogarca and in 2015, she graduated her postdoctoral studies on *Ethics Management in the Public Sector of Romania.* Silvia teaches Management, Business Ethics, Public Marketing and Creative Writing. During the last years, she published more than 40 articles in national and international journals or in the proceedings of international conferences, published two books and a chapter on *Corporate Social Responsibility in the Romanian Public Sector* in the book *Corporate Social Responsibility in Times of Crisis* (Springer, 2017). Her research covers topics from strategic management, ethics management, public marketing, corporate social responsibility and management. She is also a reviewer for two journals (Management & Marketing, The Young Economists Journal) and a member of Eurasia Business and Economics Society.

ISO 26000: A Vital Vector in Interconnecting the Human Resource Management Standards

Cătălin Mihail Barbu, Monica Logofătu, and Camelia Olari

1 Introduction

Global economic development in the post-war period has had not only positive effects (increased life expectancy, rising living standards, important breakthroughs in science and technology, etc.), but also negative effects (environmental pollution, rising differences between the rich and the poor, increases in the world population, etc.). Social responsibility theory has emerged as a result of the fact that governments have not been able to combat these negative effects. Social responsibility promote the idea that all the actors involved in the economic process (organizations, employees, customers, suppliers, public administration, civil society, local community, etc.), commonly referred to as stakeholders, must also act for the benefit of society as a whole, not just for its own benefit. These actions can be supported passively (by avoiding involvement in actions that have a negative impact on the natural and social environment as well as on society as a whole) or actively (pursuing activities aimed at achieving social goals).

Corporate social responsibility is becoming more and more important as a result of changes in values (focusing on sustainability and combating the effects of pollution, respect for human rights, guiding good practice and business practices, etc.) and consumer demands to make organizations accountable for the consequences of their actions. These changes, as well as raising awareness among all stakeholders, generate greater involvement in organizations' social responsibility initiatives, in their attempt to increase legitimacy in front of stakeholders (Sen and Cowley 2013). This legitimacy cannot be achieved unless the organization manages to communicate its actions that are socially responsible to the stakeholders.

C. M. Barbu (✉) · M. Logofătu · C. Olari
Faculty of Economics and Business Administration, University of Craiova, Craiova, Romania

S. O. Idowu et al. (eds.), *ISO 26000 - A Standardized View on Corporate Social Responsibility*, CSR, Sustainability, Ethics & Governance,
https://doi.org/10.1007/978-3-319-92651-3_3

Given that employees are a key factor in organizational success (Zadek et al. 2001), socially responsible human resources management has become a crucial issue for companies (Lopez-Cabrales et al. 2006). As Baraibar-Diez and Luna-Sotorrío (2012) point out, it is not only important to develop integrated socially responsible human resources management models but also to communicate actions in this area to employees and other stakeholders to ensure a high level of awareness.

Regarding the implementation of responsible behavior in the area of human resource management, it will depend, among other things, on the institutional characteristics of the country in which the organization operates (Diaz-Carrion et al. 2017). Given that organizations are strongly influenced by the actions of the state in which they operate and the characteristics of national cultures, particular types of capitalism generated by different cultures determine various ways of managing company resources, including human resources (Sparrow et al. 1994). By comparing the human resource management implemented in Europe and the United States, European countries see a greater influence on the institutional factor, with a higher degree of regulation. Morley (2004) shows that human resources management in Europe is affected by various institutional market forces such as government intervention, legislation, trade unions, social dialogue, while the interests of shareholders in the United States are a priority. Diaz-Carrion et al. (2017) considers state regulations to become important coercive pressures that influence the responsible social behavior of organizations. A middle way between the institutional perspective specific to Europe and the liberal perspective specific to the Anglo-Saxon space is the use by organizations of generally accepted standards in the area of human resource management. The International Standardization Organization (ISO) has taken steps since 2016 in this direction by publishing standards, specifications and technical reports in the area of human resource management.

The success of the implementation of policies, practices and programs in the area of social responsibility is influenced by employee involvement and commitment, which emphasizes the crucial role of human resources management in implementing a responsible behavior of the organization (Berber et al. 2014). In order to harmonize the organizational values with the individual ones, the organization must recruit, select and employ workers who identify themselves with organizational principles and values, modify reward systems in such a way as to encourage those behaviors that lead to improvement of the social performance of the organization, to set up employee training and development systems that are compatible with the values and principles of sustainability (Orlitzky and Swanson 2006).

The present chapter aims to undertake an exploratory research among the international human resource management standards proposed by ISO. We seek to integrate them into a management system whose backbone is ISO 26000:2010, given that this standard has an important component of human resources and that it is already implemented by many organizations. Given that human resource standards are relatively new and have not benefited from intense popularization, and the fact that research in the field is scarce, we believe that this chapter can be a relevant contribution to the field. Such an integrated management system will help human resources departments improve their performance and, ultimately, improve the performance of the organization.

The chapter is structured in five sections. The second section presents the relationship between human resource management and social responsibility. The third section provides an overview of ISO standards in the area of human resource management. In the fourth section we aim to create an integrated model of human resource management using the ISO 26000:2010 as a vector. Section 5 concludes and indicates the direction of future research.

2 The Relationship Between Human Resource Management and Social Responsibility

Since the emergence and spread of social responsibility principles, a number of researchers have investigated over time the relationship between human resource management and social responsibility of organizations (Zappala and Cronin 2002; Orlitzky and Swanson 2006; Aguilera et al. 2007; Preuss et al. 2009; Sharma et al. 2009; Maksimainen and Saariluoma 2010; Inyang et al. 2011; Buciuniene and Kazlauskaite 2012; Baraibar-Diez and Luna-Sotorrío 2012; Lis 2012; Berber 2013; Berber et al. 2014; Diaz-Carrion et al. 2017). The perception of job candidates about current practices in the area of social responsibility determines the attractiveness of the organization (Lis 2012). An organization perceived as having a responsible behavior will have good relationships with employees, which will allow the organization to gain additional benefits, including improving public image, increasing employee morale, community support, and overall performance (Zappala and Cronin 2002). Aguilera et al. (2007) shows that human resources should play a major role in the operationalization of the actions in the area of social responsibility. The combined effects of successful policies in the area of human resources management and social responsibility support the achievement of desirable behaviors, with a major contribution to the creation of an organizational climate that can sustain viable performance (Sharma et al. 2009).

Employees are one of the main internal stakeholders in the design and implementation of organizational strategies. Therefore, employee satisfaction and the allocation of a significant part of the added value by the organization is an essential element in designing social responsibility strategies and organizational investment in the broad sense (Klimkiewicz and Beck-Krala 2015). In this respect, it is very important for HRM to take on the challenge of introducing quantifiable indicators in the area of social responsibility that allow for sustainable development of the organization (Barrena-Martínez et al. 2017). At international level, there are a number of standards (some of which provide for indicators) that seek to standardize and assess social responsible behavior in HRM area, such as GRI (Global Reporting Initiative) and ISO 26000:2010.

ISO 26000:2010 provides guidance on basic principles of social responsibility, recognition of social responsibility and stakeholder involvement, core subjects of social responsibility, and how socially responsible behavior can be integrated into the organization. This international standard emphasizes the importance of results and improvements in social responsibility performance. It is recommended for all types of organizations regardless of sector, size or location. This standard supports organizations to contribute to sustainable development. Its purpose is to encourage them to go beyond compliance with the law, recognizing that compliance with the law is a fundamental duty of any organization and an essential part of their social responsibility.

Although information on socially responsible human resource management can be expressed in several different ways, Diaz-Carrion et al. (2017) found that there was a widespread use of GRI to communicate this information. The GRI Reporting Framework addresses all aspects of organizational sustainability, including strategy, governance, ethics and the economic, social and environmental aspects of the business. Although human resource management is just a dimension of sustainability, the GRI framework includes many references to policies and human resource performance. Although this standard cannot be certified, there is an upward trend in presenting the human resource indicators provided by this standard, which leads to an increasingly standardization of information disclosure in area of social responsibility in Europe.

The emergence of new ISO standards in the human resources area has introduced a number of new indicators that can be harmonized by a full integration into a socially responsible human resources management model. The tangible results of sustainable human resource management include not only employees' commitment to achieve the organization's overall goals, but also measurable contributions to human resource performance, including lowering employee exit rates, low absenteeism, improved living standards of the employees and a general increase of employee engagement, motivation and overall productivity.

3 ISO Standards in the Area of Human Resource Management

The greatest competitive advantage and the most important resource of an organization is human capital. Therefore organizations that put the human resource in the center of operations and strategies are the most effective. In the years 2016 and 2017, ISO has proposed to organizations, regardless of the sector of activity and ownership, a new set of standards that seek to improve human resource planning, recruitment of best talent, and improve organizational governance through efficient processes in the area of human resource management.

In a world where corporations tend to capture most of the economy, where the shareholder is often ranked first among the stakeholder categories, organizations do not pay attention to human resource values, they do not make reliable decisions and

Table 1 Terms used in human resource management ISO standards

General terms	Organization
	Human resources and planning
	People and organization
	Human resource metrics
	Human governance
	Sustainable employability
	Diversity and inclusion
Specific topics	Workforce planning
	Recruitment
	Workforce mobility
	Talent management
	Knowledge management

Source: Developed by the authors after ISO (2016a)

do not engage in a responsible behavior, this affecting the way of doing business and finally the profitability of the organization.

Studies show that a human resources department that performs well comparing to the human resources departments of competing firms can be related to the growing economic performance of the organization and plays a key role in the operationalization of organizational values, in improving organizational culture, in increasing the overall social responsibility of the organization. The new range of international standards proposed by ISO in human resource management aims to help human resources departments improve their performance and, ultimately, improve the performance of their organization (Naden 2016).

ISO 30400:2016 Human resource management—Vocabulary has been developed to achieve a concept harmonization in area of human resource and to maintain consistency in the vocabulary to be used in the standards in this area. This standard was created in view of the multitude of meanings and definitions of various concepts in the area of human resource management. Users of this standard are primarily practitioners and professionals in the area of human resources, students, standards developers, but also the standard address the stakeholders of all organizations, regardless of size and type of organization, especially trade unions, other employees representatives and employers' representatives (ISO 2016a). Within this standard, there are defined the human resource concepts grouped into 12 categories (Table 1).

ISO 30405:2016 Human resource management—Guidelines on recruitment. Recruitment is an important part of human resources management including identifying, attracting, selecting and hiring the most talented people. ISO 30405:2016 is a support for organizations in achieving performance in recruitment processes by providing guidance on the procedures. ISO highlights the fundamental role of the employer's reputation in recruitment by showing that there are four critical processes that influence the quality of recruitment: identification, attraction, selection and hiring. The quality of recruitment influences the results of human resources management and ultimately the performance of the organization.

Table 2 Main components of ISO 30405:2016

Processes, procedures and criteria used in recruitment	General provisions
	Potential talent pool
	Talent pool
	Applicant pool
	Candidates for further assessment
	Candidate management
	Pre-boarding
	Boarding
Guidelines for assessing the effectiveness, efficiency and impact of the recruitment process	General provisions
	Measurement process
	Analyze results
	Improvement

Source: Developed by the authors after ISO (2016b)

The standard provides guidance to those responsible for recruitment and is structured into two parts: one describes the processes, procedures and criteria used in recruitment (Clause 4), while the second part (Clause 5) provides guidelines for assessing the effectiveness, efficiency and impact of the recruitment process Table 2. The standard also includes four annexes that show the indicators and tools used in a typical recruitment processes.

If in large organizations, recruiting is done by the human resources specialists or external recruitment consultants, in small organizations, recruitment is usually carried out by people from the organization management without the necessary skills in the area of human resources management. The standard provides guidance for all recruiting persons regardless of formal training in the field.

ISO 30409:2016 Human resource management—Workforce planning has been developed by ISO to support organizations of any size, industry or sector in their workforce planning activities in such a way as to fully adapt to the requirements of the labor market and the environment increasingly complex competition. The standard is structured into five parts (Table 3). The standard also includes an appendix which expose a typical planning process within a small organization.

ISO 30408:2016 Human resource management—Guidelines on human governance. Governance of an organization is the system through which the organization is managed, controlled and accountable. Governance establishes rules, systems, processes and behaviors that implemented contribute to the organization's work, value creation in a responsible way. Employees are at the heart of each organization's work, representing a vital and ubiquitous stakeholder in all organizations. Human governance is the system whereby employees within an organization are led and held accountable, promoting appropriate behavior within an organization (Fig. 1).

By aligning human governance practices to overall organizational governance practices, the organization can anticipate and manage human resource risks, manage human capital costs and evaluate investments made in it, promote participatory

Table 3 Main components of ISO 30409:2016

Value of workforce planning	General provisions
	Relevance of workforce planning to different stakeholders
	Value to organizations
	Value to industries or sectors
	Value to regions
	Value to the public sector
	Value to people
	External and internal influences
Structure of workforce planning	General provisions
	Structure of workforce planning
	Process of workforce planning
Risk management	General provisions
	Workforce planning checklist
	Start of planning
	Analyze the data
Key components of workforce planning	Identify risk management strategies, action plan and report
	Secure the resources
	Develop and implement an action plan
Implementation	Develop a mechanism for monitoring and measuring workforce planning outcomes
	Integrate strategic workforce planning into core business planning process

Source: Developed by the authors after ISO (2016d)

management and a high degree of commitment to the organization's activities by employees, including human responsibility activities, all of which resulting in optimize of the overall organization performance (ISO 2016c).

ISO 30408:2016 provides guidance on the tools, processes and practices needed to establish, maintain and continuously improve the human governance of the organization, regardless of size, industry or sector, by ensuring that human governance practices are aligned with the practices of overall organizational governance. The structure of the standard is sequential, starting from the establishing of the human governance principles and ending with the implementation.

ISO/TS 30407:2017 Human resource management—Cost-Per-Hire. Cost-Per-Hire (CPH) represents a measurement tool from the category of technical specifications that allows the evaluation of the efficiency of the recruitment, selection and hiring process within an organization. The information provided by using this metric helps the recruitment of human resources in terms of budgeting and comparing recruitment costs. This document covers the definition, assembly and techniques documentation necessary for calculation of CPH. ISO/TS 30407:2017 allows calculation of recruitment costs and ensures comparability by providing an algorithm for determining recruitment unit costs.

Starting from the fact that organizations have different characteristics in the recruitment, selection and employment process, ISO/TS 30407:2017 provides an

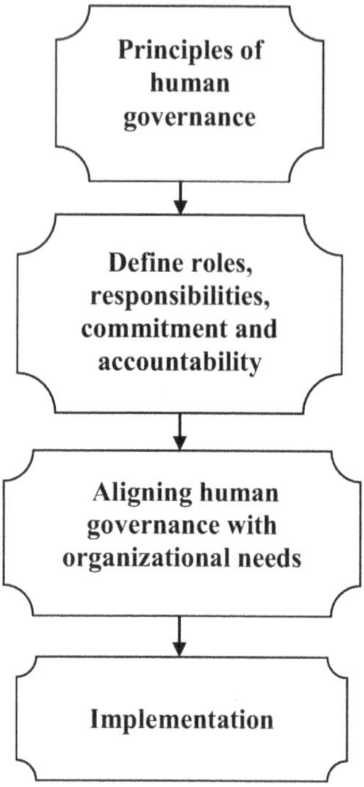

Fig. 1 Structure of ISO 30408:2016. Source: Developed by the authors after ISO (2016c)

adaptive methodology that allows, the recruitment cost to be tailored to the specific organization, while allowing comparisons with other organizations. The structure of the standard starts from internal costs and ends with assembling the various cost components (Fig. 2).

Cost-Per-Hire Internal (CPHI) has only internal use to the organization being a more complex expression of recruitment costs. Cost-Per-Hire Comparable (CPHC) ensures comparability with other organizations (measure the efficiency of recruitment compared to other organizations). Hire-Cost-Ratio (HCR) is a measure of the cost-effectiveness of recruiting by comparing the total cost of recruitment with the employee's total wage in the first year of employment.

ISO/TS 30407:2017 has the role of maintaining the quality and transparency of recruitment, ensuring the reduction of computational errors and facilitating regular audits of recruitment processes.

ISO/TR 30406:2017 Human resource management—Sustainable employability management for organizations is a technical report that focuses on sustainability in the area of human resource management. Sustainability of employment under ISO/TR 30406:2017 is seen both from the employee's point of view (the opportunity to work in a satisfactory environment to ensure a quality of working life) and the employer (the commitment of an organization to meet the needs of employees, to ensure their well-being and the balance between individual needs and organizational requirements). Practically, employment sustainability involves assessing labor market requirements, organizational needs assessment and planning, retraining and

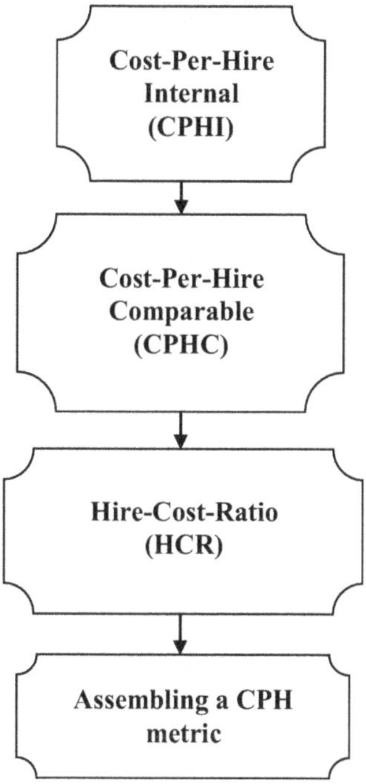

Fig. 2 Structure of ISO 30407:2017. Source: Developed by the authors after ISO (2017b)

Table 4 Main components of ISO/TR 30406:2017 and stages of implementation process

Components	Stages of implementation process
Sustainable employability on different levels	Planning and design
Guiding principles	Measuring and disseminating the sustainable employability tool
Cluster 1: Integration of sustainable employability into policy	Analyzing sustainable employability factors and preparing report
Cluster 2: Work capabilities, content, complexity, relationships and conditions	Meeting of key stakeholders
Cluster 3: Vitality and health	Implementation

Source: Developed by the authors after ISO (2017a)

redistribution of staff in case of structural changes, ensuring the balance between job requirements and employee skills. ISO/TR 30406:2017 is a bridge between human resources policy and organization general policy, quality policy, and quality improvement cycle.

In this technical report are defined the levels and principles of human resource sustainability, describing the main concepts grouped on 3 clusters Table 4. The annex describes a process of implementing policies on sustainability of human resources employment.

ISO 10015:1999 Quality management—Guidelines for training provide guidance to organizations and their employees to better address training issues. This

Table 5 Main components of ISO 10015:1999 Quality management—guidelines for training

Components	Stages of implementation process
Guidelines for training	Stage of training
	Defining training needs
	Designing and planning training
	Providing for the training
	Evaluating training outcomes
Monitoring and improving the training process	General provisions
	Validation of the training process

Source: Developed by the authors after ISO (1999)

international standard provides guidelines that complement and details the references in the ISO 9000 family standards on training. The area of training is the first human resource area for which ISO has released a standard. The quality of products and services cannot be achieved if employees are not trained to meet customer requirements.

The role of this international standard is to provide guidance that can help an organization identify and analyze training needs, design and plan training, provide training, assess training outcomes, monitor and improve the training process, and achieves the goals. ISO 10015:1999 emphasizes the contribution of training to continuous improvement and demonstrates that training is an investment in people that bring profit to the organization. The standard is structured in two parts, focusing on training guidelines and ways to monitor and improve training processes Table 5.

Together with the standards, technical specifications, technical reports released until the present, the ISO/TC 260 Human Resources Management Committee is developing other human resource management documents to be published in the coming years:

- ISO/DIS 30401 Knowledge management systems—Requirements
- ISO/DTS 30410 Human Resource Management—Impact of Hire Metric
- ISO/DTS 30411 Human Resource Management—Quality of Hire
- ISO/DIS 30414 Human resource management—Guidelines for human capital reporting for internal and external stakeholders
- ISO/AWI 30415 Diversity and inclusion
- ISO/AWI 30416 Human resource management—Workforce management
- ISO/AWI 30419 Guidelines for ensuring a positive candidate experience during the recruitment process.

4 Designing an Integrated Model of Human Resource Management

The design of an integrated model of human resource management can generate a harmonization of HR processes. If multinationals would adopt this model and require suppliers to do the same, the approach of human resources within an organization will be done in a way that will achieve sustainable performance. Indicators that characterize human resources would be comparable if they were calculated using the same generally accepted model (Bocean and Sitnikov 2015). Human capital is the most important resource of any organization, and therefore the use of best practices in human resource management will be a competitive advantage for any organization.

In their work, Martín-Alcázar et al. (2005) propose a sequential and structured model of integration of socially responsible human resources management that has the following characteristics: universality, contingency, configurational and contextual (Barrena-Martínez et al. 2017).

First of all, the universalist perspective of an integrated model of human resource management is based on the premise that there are universally successful models used in human resource management. Consequently, in order to ensure a universal character, it is necessary to identify those effective human resources policies and practices that do not differ according to context, country or other internal or external variables to organizations. Taking into account the complexity of the environment in which organizations operate, it can be seen that there are a large number of external variables that can influence the specificities of human resource practices. ISO standards in the area of human resources, and especially the ISO 26000 standard, have managed to overcome this difficulty, these standards having a universal character.

However, the complexity of the environment can be countered by an integrated model of socially responsible human resources management through a contingent perspective that can provide a better explanation of the effects and interconnections between human resources policies and the various internal variables (domain, structure, size, technology, etc.) or external variables (acting in the organizational environment), with the aim of increasing accountability.

The third, configurational feature allows the definition of a coherent system of socially responsible human resources by configuring the components of the model according to the interactions of these policies with the internal and external variables. This adjustment can lead to the development of behavioral models in the area of human resource management, thus contributing to improving organizational performance. Although the configurational feature allows a flexible approach to the effects and interactions of the human resources system with internal and external variables, institutional pressures and stakeholder requirements should also be taken into account in order to better understand the context.

The contextual feature would determine the adaptability of the model to the contextual aspects: the socio-economic context (political, institutional, social, legal, economic and environmental components, national culture, education) and the organizational context (dimension, technology, strategy, organizational culture).

The integrated model of human resource management (IMHRM) that we propose is based on ISO 26000:2010 as an integrator vector that allows aggregation and harmonization of ISO standards, specifications and technical reports in the area of human resources.

Among the main topics that an organization should address in setting its social responsibility priorities are labor practices. An organization's labor practices include all principles, policies, and actions related to human resources within the organization. Issues covered by ISO 26000:2010 in the core subject of labor practices are as follows:

- employment and employment relationships,
- conditions of work and social protection,
- social dialogue,
- health and safety at work,
- human development and training.

IMHRM proposes that, the issues set by ISO 26000:2010 in the core subject of work practices, to be the foundation of the integrated system through the defined actions and expectations. Within each human resource management area defined by ISO 26000:2010 are integrated the provisions and clauses of the various standards, specifications and technical reports published by ISO in the area of human resources (Fig. 3).

The model can be further supplemented with other standards, technical specifications, and technical reports which will be published later.

The IMHRM model is based on Jorgensen's integrated management system methodology (Jogersen 2008), proposing three important levels of integration (compatibility, coordination, integration). Compatibility involves ensuring the correspondence between the ISO 26000 clauses and the clauses of ISO human resource standards, eliminating duplication, confusion and redundant activities. Coordination ensures alignment of policies and objectives within processes. Integration involves interweaving policies and objectives, as well as developing a unified process underpinning effective human resource management.

The IMHRM model should be based on a process-based approach involving the use of the PDCA continuous improvement cycle. The planning phase allows the initial analysis, the establishment of some indicators for the measurement of the human resources performance. The implementation phase involves the implementation of all the actions set out in the organization's current activities. The verification phase consists of assessing the impact of the actions in order to determine the level of achievement of the objectives set in the area of human resources. The action phase requires the organization's human resources plans to be corrected in order to achieve the proposed objectives.

Such an integrated model will allow a more efficient management of human resources, which will provide the organization with a higher value, given that its employees are the main vector in creating value.

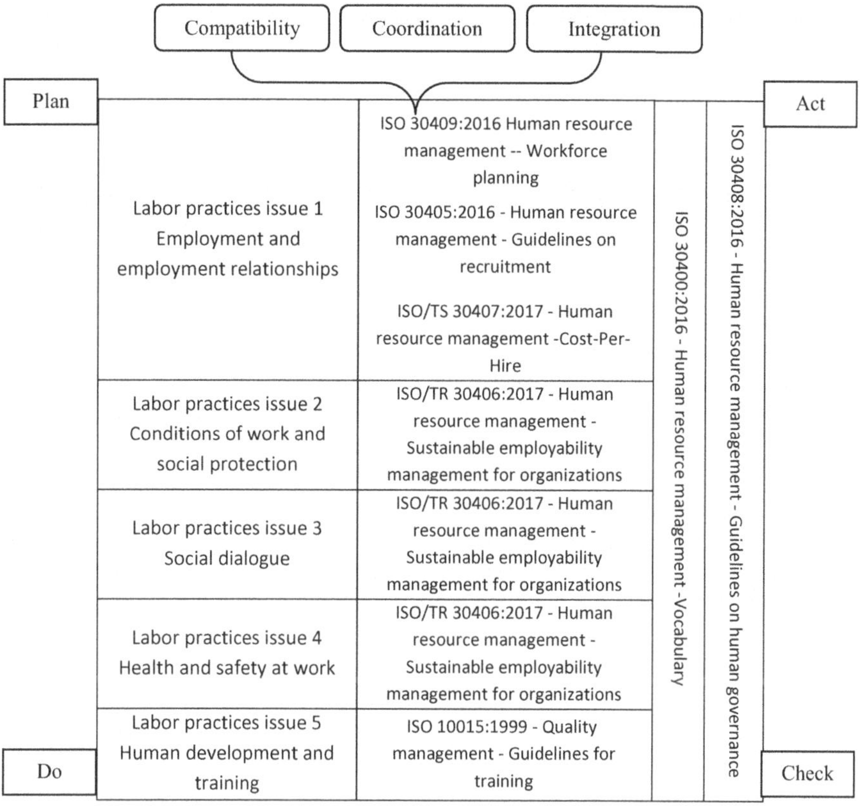

Fig. 3 Integrated model of human resource management (IMHRM). Source: Developed by the authors

The field of human resources has not been so far subject to standardization initiatives due to active opposition from the management of organizations. Top managers consider the human resources to contribute to the competitive advantage of the organization and they do not want other organizations to be able to replicate their human resources management. However, in recent years, managers have become aware that a soft standardization of the human resources management can bring important benefits to the organization. The model proposed in this chapter integrates human resource standards and allows managers to use its specifications and adapt them to their organization, being a theoretical landmark that can be applied for pragmatic purposes.

5 Conclusions and Further Research

Although the area of social responsibility has received particular attention in recent years, it can be seen that there is no clear and unanimously accepted definition of each socially responsible policy. In addition, within the sphere of social responsibility there are different views on the importance of the different subjects. This disagreement is generated by the ambiguity and variety of CSR actions, which go beyond legal frameworks, being in line with business ethics, which generates a multitude of opinions influenced by national and organizational cultures. Since employees are a key factor in organizational success, human resource management has become a crucial social issue for organizations.

Organizations are aware that the sustainability of a business is influenced by their competitiveness, reputation and ability to attract and retain talented human resources. Taking into account the impact of their economic, social and environmental activities, sustainable organizations are trying to get the support of a broad and diverse group of stakeholders—both internal and external—to operationalize their business strategies. Human resource management can play a key role in implementing the sustainability strategy. Using human resource abilities in the organizational process, change management and culture management, human resources management can help create and implement a sustainable business strategy across the organization.

In the area of human resources, due to differences in the type of organization, sector, size or location, labor practices are homogeneous. ISO 26000:2010 has introduced labor practices among its core subjects, setting expectations and actions that organizations must undertake in this area to prove themselves socially responsible. These are just general guidelines, insufficient to achieve efficient human resource management. Since 2016, ISO has released standards, technical specifications, and technical reports in the area of human resources. These documents provide guidance to organizations in different areas of human resource management. Within this chapter, we propose an integrated model of human resource management (IMHRM) based on ISO 26000:2010 as an integrator vector that allows aggregation and harmonization of the various standards, technical specifications and technical reports published by ISO in the area of human resources. The model will be developed in future papers and will be further supplemented with other standards, technical specifications and technical reports which will be published by ISO so as to provide an optimal tool for organizations to achieve effective and social responsible human resource management.

The integrated model of human resource management which we are proposing is trying to add a new social nuance to the management of human capital. In this respect, it is necessary that the human resources policies aim at not only higher individual employees' performance but also balancing the personal and professional expectations of the employees, increasing their standard of living, loyalty to the organization and full involvement in achieving organizations' goals.

The model is addressed to all types of organizations, regardless of sector, size, or location. From the academic point of view, the universalists' perspective is the most relevant approach to a human resource management model because it provides a simpler way of understanding the relationship between human resource management practices, giving all organizations whatever the characteristics and context in which they act the occasion to approach human resources in a responsible way.

References

Aguilera, R. V., Rupp, D. E., Williams, C. A., & Ganapathi, J. (2007). Putting the S back in corporate social responsibility: A multilevel theory of social change in organizations. *Academy of Management Review, 32*(3), 836–863. https://doi.org/10.5465/AMR.2007.25275678

Baraibar-Diez, E., & Luna-Sotorrío, L. (2012). Transparencia social e hipótesis del impacto social. Análisis en el IBEX35/Social transparency and social impact hypothesis. Analysis on the IBEX35. *Universia Business Review, 36*, 108–123.

Barrena-Martínez, J., López-Fernández, M., & Romero-Fernández, P. M. (2017). Socially responsible human resource policies and practices: Academic and professional validation. *European Research on Management and Business Economics, 23*, 55–61.

Berber, N. (2013). *Linkage between human resource management and corporate social responsibility*. In 18th international scientific conference SM 2013, Subotica: Faculty of Economics, pp. 1024–1036.

Berber, N., Stangl Susnjar, G., Slavic, A., & Baosic, M. (2014). Relationship between corporate social responsibility and human resource management - as new management concepts - in central and Eastern Europe. *Engineering Economics, 25*(3), 360–369.

Bocean, C. G., & Sitnikov, C. S. (2015). Setting international standards in human resources. *Management and Marketing, XIII*(1), 16–22.

Buciuniene, I., & Kazlauskaite, R. (2012). The linkage between HRM, CSR and performance outcomes. *Baltic Journal of Management, 7*(1), 5–24.

Diaz-Carrion, R., López-Fernández, M., & Romero-Fernandez, P. M. (2017). Social human resource management transparency in Europe: A cross-country analysis. *Universia Business Review, Second Quarter, 54*, 70–89.

International Organization for Standardization (ISO). (1999). *ISO 10015:1999 - quality management - guidelines for training*. Geneva: International Organization for Standardization. Retrieved October 23, 2017, from https://www.iso.org/obp/ui/#iso:std:21231:en

International Organization for Standardization (ISO). (2010). *ISO 26000:2010 - guidance on social responsibility*. Geneva: International Organization for Standardization. Retrieved October 23, 2017, from https://www.iso.org/obp/ui/#iso:std:iso:26000:ed-1:v1:en

International Organization for Standardization (ISO). (2016a). *ISO 30400:2016 - human resource management - vocabulary*. Geneva: International Organization for Standardization. Retrieved October 27, 2017, from https://www.iso.org/standard/66032.html

International Organization for Standardization (ISO). (2016b). *ISO 30405:2016 - human resource management - guidelines on recruitment*. Geneva: International Organization for Standardization. Retrieved October 27, 2017, from https://www.iso.org/standard/64149.html

International Organization for Standardization (ISO). (2016c). *ISO 30408:2016 - human resource management - guidelines on human governance*. Geneva: International Organization for Standardization. Retrieved October 27, 2017, from https://www.iso.org/standard/63492.html

International Organization for Standardization (ISO). (2016d). *ISO 30409:2016 human resource management - workforce planning*. Geneva: International Organization for Standardization. Retrieved October 27, 2017, from https://www.iso.org/standard/64150.html

International Organization for Standardization (ISO). (2017a). *ISO/TR 30406:2017 - human resource management - sustainable employability management for organizations*. Geneva: International Organization for Standardization. Retrieved October 27, 2017, from https://www.iso.org/standard/72327.html

International Organization for Standardization (ISO). (2017b). *ISO/TS 30407:2017 - human resource management - cost-per-hire*. Geneva: International Organization for Standardization. Retrieved October 27, 2017, from https://www.iso.org/standard/62975.html

Inyang, B. J., Awa, H. O., & Enuoh, R. O. (2011). CSR-HRM Nexus: Defining the role engagement of the human resources professionals. *International Journal of Business and Social Science, 2*, 118–126.

Jogersen, T. H. (2008). Towards more sustainable management systems: Through life cycle management and integration. *Journal of Cleaner Production, 16*(10), 1071–1080.

Klimkiewicz, K., & Beck-Krala, E. (2015). Responsible rewarding systems - the first step to explore the research area. In *Prace Naukowe Uniwersytetu Ekonomicznego we Wroclawiu*. Social Responsibility of Organizations Directions of Changes (pp. 66–79). Wrocław: Wrocław University of Economics.

Lis, B. (2012). The relevance of corporate social responsibility for a sustainable human resource management: An analysis of organizational attractiveness as a determinant in Employees' selection of a (potential) employer. *Management Revue - The International Review of Management Studies, 23*(3), 279–295.

Lopez-Cabrales, A., Valle, R., & Herrero, I. (2006). The contribution of core employees to organizational capabilities and efficiency. *Human Resource Management, 45*(1), 81–109.

Maksimainen, J., & Saariluoma, P. (2010). How human resource management and human capital management influence corporate social responsibility (CSR). *International Journal of Knowledge, Culture and Change Management, 10*(5), 111–126.

Martín-Alcázar, F., Romero-Fernández, P. M., & Sánchez-Gardey, G. (2005). Strategic human resource management: Integrating the universalistic, contingent, configurational and contextual perspectives. *The International Journal of Human Resource Management, 16*(5), 633–659.

Morley, M. J. (2004). Contemporary debates in European human resource management: Context and content. *Human Resource Management Review, 14*(4), 353–364.

Naden, C. (2016). *Improve the bottom line with a new range of human resource management standards*. International Organization for Standardization. Retrieved September 22, 2017, from https://www.iso.org/news/2016/09/Ref2111.html

Orlitzky, M., & Swanson, D. L. (2006). *Socially responsible human resource management, human resource management ethics*. Greenwich, CT: Information Age.

Preuss, L., Haunschild, A., & Matten, D. (2009). The rise of CSR: Implications for HRM and employee representation. *The International Journal of Human Resource Management, 20*(4), 953–973.

Sen, S., & Cowley, J. (2013). The relevance of stakeholder theory and social capital theory in the context of CSR in SMEs: An Australian perspective. *Journal of Business Ethics, 118*(2), 413–427.

Sharma, S., Sharma, J., & Devi, A. (2009). Corporate social responsibility: The key role of human resources management. *Business Intelligence Journal, 2*(1), 205–213.

Sparrow, P., Schuler, R. S., & Jackson, S. E. (1994). Convergence or divergence: Human resource practices and policies for competitive advantage worldwide. *International Journal of Human Resource Management, 5*(2), 267–299.

Zadek, S., Hojensgard, N., & Raynard, P. (2001). *Perspectives on the new economy of corporate citizenship*. Copenhagen: Copenhagen Centre.

Zappala, G., & Cronin, C. (2002). The *employee dimensions of corporate community involvement in Australia: Trends and prospects*. In 6th ANZTSR conference, 27–29 November, Auckland, pp. 1–24.

Cătălin Mihail Barbu, is an Associate Professor at the Department of Management, Marketing, Business Administration, in Faculty of Economics and Business Administration within the University of Craiova, Romania. In 2006 he obtained his PhD degree in International Economic Relations within the Faculty of Economics, University of Craiova, Romania. Since 2001 he has been teaching at Faculty of Economics and Business Administration, University of Craiova taking courses on topics such as Marketing, International Marketing, Brand Management. He has been co-operating in research projects with national and international universities and organizations. He is a member of the European Marketing Academy and he serves on the Editorial board of several journals in Romania and abroad.

Monica Logofătu, (Mirea) is a PhD student in management area at the Faculty of Economics and Business Administration at the University of Craiova, Romania. In 2006 she graduated with a Bachelor's Degree in Geography and French Language at University of Craiova. In 2017, she graduated with a Master's program in Public Organizations Management at the Faculty of Economics and Business Administration, University of Craiova. Her areas of interest are in human resource management, education, social responsibility.

Camelia Olari, is currently a Retail Group Manager at BCR and worked in the Banking sector for 9 years, mainly in Human Resources Management projects—selecting, recruiting, training, mentoring, professional development and employees' certification. She started her professional path in Citibank as a Direct Sales Agent in 2004 and since 2008 she has filled management positions in banking and sales areas. Currently she is part of the Professional Certification Program of the Employees as an Internal Evaluator of Professional staff. This experience has helped her to discover and develop certain abilities, concentrated on the Human Resource Quality (as she is currently a PhD student in Management, in the Faculty of Economics and Business Administration, University of Craiova, Romania, with a focus in Human Resources Quality Management).

The Adoption of ISO 26000 in Practice: Empirical Results from The Netherlands

Lars Moratis and Alice Tatang Widjaja

1 Introduction

While literature on the diffusion and the adoption of standards by organizations is vast (see for instance Brunsson and Jacobsson 2000; Corbett and Kirsch 2001; Guler et al. 2002; Delmas 2002; Neumayer and Perkins 2005; Christmann and Taylor 2006; Mueller et al. 2009; Del Mar Alonso-Almeida and Rodriguez-Anton 2011), studies into the adoption of CSR-related standards and especially comprehensive CSR standards have been rather scarce. This observation applies even more when one looks at the availability of empirical studies into this subject. The bulk of the research has focused on management system standards that relate to CSR *issues*, especially environmental management standards (e.g., ISO 14001) and, to a lesser extent, social management systems (e.g., SA8000). These studies have addressed different aspects such as the motives, benefits, impact and institutional factors of standards adoption.

Launched in November 2010, the ISO 26000 standard is the most recent CSR[1] standardization initiative, aiming to provide *"guidance on the underlying principles of social responsibility, recognizing social responsibility and engaging stakeholders,*

[1]ISO 26000 uses the term 'social responsibility' which it abbreviates with SR. The primary reason for omitting the word 'corporate' from the commonly used term CSR lies in the fact that the standard claims it applies not only to corporations, but to organizations of all sorts, sectors, and

L. Moratis (✉)
Antwerp Management School, Antwerp, Belgium

NHTV Breda University of Applied Sciences, Breda, Netherlands
e-mail: lars.moratis@ams.ac.be

A. T. Widjaja
Evellia Company, Paris, France
e-mail: alice@evellia.com

S. O. Idowu et al. (eds.), *ISO 26000 - A Standardized View on Corporate Social Responsibility*, CSR, Sustainability, Ethics & Governance,
https://doi.org/10.1007/978-3-319-92651-3_4

the core subjects and issues pertaining to social responsibility and on ways to integrate socially responsible behaviour into the organization" (ISO 2010: vi). The standard was developed in a global multi-stakeholder approach led by the International Organization for Standardization (ISO), which involved the consultation of hundreds of CSR experts and organizations worldwide. ISO 26000 reflects international consensus on the meaning of CSR and on the CSR issues organizations of all sorts, sectors, and sizes should address in their operations. As an example of a transnational norm-building network (Mueckenberger and Jastram 2010), the standard possesses a high degree of output legitimacy (Mena and Palazzo 2012). As opposed to well-known ISO standards such as ISO 14001 and ISO 9001, ISO 26000 does not include requirements for a management system but merely provides organizations with guidance. Perhaps the most striking and debated aspect of the standard is that it is not intended for certification. The standard merely offers organizations guidance by providing suggestions in order for them to engage, accelerate, monitor or evaluate the implementation of CSR (e.g., Hahn 2012; Moratis and Widjaja 2014; Delchet-Crochet and Vo 2013). While the adoption of the standard is still in its early stages, ISO 26000 may contribute to the development and assessment of the roles and responsibilities of business in society.

Existing literature does not offer many predictions on the adoption of comprehensive CSR standards and it remains to be seen whether the peculiar aspects of ISO 26000 affect its adoption. This article aims to shed light on this issue by examining the determinants of organizations' adoption of comprehensive CSR standards, particularly ISO 26000. It presents the results of original exploratory empirical research conducted among corporate CSR professionals in the Netherlands. As such, it provides relevant empirical findings that complement existing literature and that have value for organizations wanting to adopt CSR standards. It also provides insights that may be relevant for standards developers and policy makers in the field of CSR that want to encourage CSR behavior by organizations.

The article is structured as follows. First, we provide an overview of relevant literature on the diffusion and adoption of CSR-related standards. We then present the results of our empirical research—which include both qualitative and quantitative findings—identifying the most important factors of adoption of ISO 26000. Finally, we draw several conclusions from our data, briefly discuss them from a theoretical viewpoint, formulate several avenues for future research and speculate on the use of the insights resulting from this article for practice.

---

sizes. In this article I have chosen to use the accepted term CSR instead of SR. In doing so, of course, I do not imply to refer solely to corporations per se.

2 Literature on the Diffusion and Adoption of CSR Standards

Literature on the diffusion and adoption of CSR standards has primarily focused on standards that address CSR issues (e.g., environment, quality, supply chain labour conditions) rather than comprehensive CSR standards. With the objective to develop a predictive model for the motivation of ISO 14001 certification, Quazi et al. (2001) identified a number of firm-specific adoption factors and tested them in a survey among Singaporean firms. Top management's concern, head office environmental practices, meeting environmental regulations, and cost savings were found to determine the firm's decision to adopt ISO 14001. However, factors beyond the firm's boundaries, such as competitors' actions and pressure from buyers, were not taken into consideration. Similarly, Curkovic et al. (2005) revealed in a qualitative study the factors that determine the successful certification for ISO 14001 by US firms. Past experience with quality management standards, organizational structure and size, percentage of sales to end-consumers, location of headquarters, and percentage of exports were found to be of influence. Even though Curkovic et al. argued regarding the decision to achieve ISO 14001 certification that "such an undertaking does not take place in a vacuum" (ibid., p. 1392), factors such as consumer demand and government regulation were mentioned as a side note only.

Regarding other anticipated benefits of a standard, Potoski and Prakash (2004) have argued that the adoption of ISO 14001 by key export destinations of companies creates incentives for them to adopt this standard. In addition, not only cost savings but also the public perception of the firms has to be taken into consideration. In a study on Swedish companies Poksinska et al. (2002) showed that marketing motives for adopting ISO 14001, such as improved corporate image, marketing advantage, and responding to customer demand, were valued over internal performance motives. Moreover, an improved relation with communities and authorities through demonstrating corporate commitment to environmental issues could be experienced by firms that have adopted standards. Beyond the motivation to enhance firm reputation and legitimacy, Zadek (1998) stated that ensuring quality and eventually supporting CSR are important reasons to implement an ethical standard. Looking at SMEs, Miles et al. (1999) reviewed the importance of adopting ISO 14001. While Drobny (1997) identified resource and internal auditor constraints, a strong entrepreneurial culture, and costs of certification as critical barriers to implement ISO 14001 in SMEs Miles et al. (1999) concluded that in industrial markets ISO 14001 certification might become a necessity for SMEs due to the supplier selection of multinational enterprises.

With regard to the business context of firms, Delmas (2002) compared the regulatory, normative and cognitive aspects of the US and European institutional environment and their impact on the costs and revenues of ISO 14001 adoption. The study showed that the early adoption of ISO 14001 in Europe was enhanced by the institutional environment through lowering transaction costs and highlighting perceived benefits of the adoption, whereas in the US the institutional environment did not support or demand the adoption. In a more recent study, Delmas and Montes-Sancho (2011) have analyzed the effect of national institutional factors on the adoption of ISO 14001, distinguishing between early and later phases of standard adoption. In the early phase, the authors say, regulative and normative forces within the institutional environment may work against each other. Regulative or coercive forces play a relatively more important role in the early stages of standards adoption, while normative forces (e.g., the diffusion of other management standards) and trade-related factors appear more important in later phases.

Using a sample of US firms that had certified to ISO 14001, Bansal and Hunter (2003) tested two possible explanations for the early adoption of this standard. They hypothesized that firms either seek to reinforce their current environmental strategy and gain a first-mover advantage through early certification, or alternatively seek to reorient their strategy and gain practical and operational benefits associated with the implementation of the standard. It was found that reinforcement of a firm's environmental legitimacy and its international presence rather than strategic reorientation were the main motive for early adoption of ISO 14001. Moreover, they suggested that widespread adoption of ISO 14001 depends on whether industry leaders will adopt the standard, whether the standard is perceived to provide a firm with legitimacy, and whether it is well-known and therefore easily recognizable by stakeholders. Similarly, in a panel study on 102 countries, Delmas (2003) showed that firms seek both efficiency through cost minimization and legitimization of the institutional environment when adopting ISO 14001. Also, the study revealed that experience with previous standards such as ISO 9001 had a positive impact on the adoption of a new standard. This could be explained by lower implementation costs and better information availability.

Research that has focused specifically on the adoption of comprehensive CSR standards has been rather limited, however. Results from studies into environmental management standards may be applicable to CSR standards as well, due to the similarities between the two (Delmas and Montes-Sancho 2011). Nikolaeva and Bicho (2011) studied the voluntary adoption by companies of the GRI (Global Reporting Initiative) guidelines, which is the most widely used standard for sustainability reporting. The study revealed that the adoption of GRI was mainly used as a management tool to improve firms' reputation and to gain legitimacy. Referring to the institutional environment, competitive and media pressure in relation to a firm's own CSR approach were found to be the main drivers for adoption. Similarly, Miles and Munilla (2004) discussed the impact of the adoption of the SA8000 standard on marketing. They propose that a firm's adoption of SA8000 will lead to product differentiation in consumer and industrial markets, and therefore to a competitive advantage in terms of reputation.

In addition to or as a substitute for adhering CSR standards, firms have also been developing their own codes of conduct with respect to CSR. According to Levis (2006), multinational companies increasingly adopt CSR codes in order to respond to stakeholders' demand for responsible business and to avoid regulation by public authorities. From an internal firm perspective, the motivation for adopting CSR codes lies in the specific organizational values and the desired reputation of the firm. Furthermore, pressure by peers and partners can lead to adoption, for instance when external auditors want to promote their environmental audit practices (cf. Leipziger 2010). However, Levis (2006) stated that since companies develop and impose their own codes of conduct, validity and independency cannot be assured. Mueller et al. (2009) contend that the different requirement levels of the variety of available standards may lead companies to prefer working with those with low exigencies, thereby using them as a legitimacy front. As a consequence, the authors say *"This strategy jeopardizes the reputation of social and environmental standards among stakeholders and their long-term trust in these instruments of CSR, meaning that all expenses for their implementation are of no avail for the companies"* (p. 509).

Especially the work of Castka and Balzarova is worth looking at in more detail here as these authors have been among a few who have focused their scholarly efforts on the adoption and diffusion of ISO 26000. In one of their articles, Castka and Balzarova (2008b) focused on how certification will impact the uptake of CSR standards. They argued that it would be preferable for the adoption of ISO 26000 if it would not be a certification standard. Even though certification may support the global diffusion of ISO standards in general, they dismiss this option due to important drawbacks of certification, pointing at the *"inconclusiveness in findings whether adopters actually do outperform non-adopters, an undesirable focus on compliance rather than on performance in many organizations and using certification to raise trade barriers and execute power in global networks"* (p. 240). In a second article, the authors compared ISO 9001 and ISO 26000 and argue that ISO 9001 may provide a structural and infrastructural platform for organisations to develop and adopt CSR (Castka and Balzarova 2008a). They also plea for ISO 26000 to facilitate a shift from customers' to a stakeholders' focus, hence creating a business-to-society orientation in organizations. In a third article that deals with ISO 26000 and supply chains, the authors formulated a series of propositions about the diffusion and adoption of the CSR standard among firms and developed a research agenda, constructing several hypotheses on the CSR orientation of organizations and networks, differences in regulatory systems, and the role of governments and national environments (Castka and Balzarova 2008c). They speculate on the diffusion of ISO 26000 among companies and formulate several predictors for this, although they, too, do not provide any empirical material to corroborate these. Drawing on the work of McWilliams et al. (2006) they look at strategic, altruistic and coercive motives by companies to adopt a CSR agenda. Their prediction is that *"only organizations that will be able to balance the cost of implementing the standard against perceived benefits from this action will choose to adopt the standard"*, that *"organizations will most likely adopt the standard if their most salient stakeholders recognize and value ISO 26000—otherwise organizations choose other means to deal with their social responsibilities"* and that they will seek legitimacy for their CSR agendas and adopt ISO 26000 if this is viewed as the proper approach (pp. 283–284). In addition, Castka and Balzarova expect local governments and other key local agents will play an important role in the diffusion of ISO 26000 (e.g., experts involved in the development process and the post-publication activities).

3 Determinants of the Adoption OF ISO 26000: Methodology and Results

Despite the aforementioned studies, which give an indication of what factors may play a role in the adoption of comprehensive CSR standards, empirical insight into the adoption of these CSR standards in general and ISO 26000 in particular still lacks. This article is based on exploratory research that the authors engaged in in the period mid-2011 to mid-2012 through which we examined the adoption of comprehensive CSR management standards. As such, this research aimed to fill the observed gap.

One of the research projects focused on the identification of determinants of the adoption of CSR standards, particularly ISO 26000. The first stage of this project was of a qualitative nature, based on interviews with 23 experts in the field of CSR and CSR standards. The results of this first empirical stage of the research have been reported on extensively in a recent paper (Moratis and Widjaja 2014). In total, 31 determinants for CSR standards adoption were identified (see Table 1), which we were able to categorize into five analytically distinctive categories. These categories were:

1. *External forces/requirements of the market*: these determinants include the attitude and actions of customers, competitors, the government, suppliers, consultants, and NGOs and their possible influence on organizations
2. *Organizational characteristics*: these determinants include the attitude of the management as well as the experience of an organization
3. *Intangible characteristics of the standard*: these determinants relate to the features that are not easily quantified or measured but are rather subjective and therefore experienced differently by organizations, such as the accessibility and name recognition of a standard, the materiality to SMEs, and the provision of structure and legitimacy to organizations
4. *Tangible characteristics of the standard*: these determinants relate to the features that are more easily quantified or measured such as the costs, the possibility of certification, the improvement of actual CSR performance and the implementation or substitution ability of the standard
5. *Characteristics/support of the standards organization*: these determinants relate to the market position, the support functions, and the legitimacy of ISO and NEN (national standardization body in the Netherlands), and the certification organizations respectively. Also, the involvement of stakeholders, the international acceptance of the standard, and the expertise of auditors were regarded as controlled or influenced by the standards organization and therefore classified as characteristics of the standards organization

Table 1 Overview of determinants of the adoption of ISO 26000 (qualitative stage)

External forces/ requirements of the market	Organizational characteristics	Intangible characteristics of the standard	Tangible characteristics of the standard	Characteristics/ support of the standards organization
The demands and wishes of customers (such as clients, buyers, the government and other customers)	The experience of an organization in working with standards	The accessibility of the standard (in terms of content, language, scope and time)	The expected implementation and certification costs	The market position (customer base, network, budget name recognition) of the organization that developed the standard
The existence of alternatives (beside standards) to demonstrate CSH performance such as benchmarks, scans, awards and reporting on CSR	The attitude of the management of a company towards the standard	Providing an organization with more structure by means of the standard	The possibility to obtain a certificate	Continuously improving the standard by the organization that developed the standard
The choice for a certain standard by an early majority of companies		Materiality to small and medium enterprises	Complementarity of the standard (Note: Complementing other or already implemented standards)	The provision of industry and sector supplements by the organization that developed the standard or by industry associations
The attitude of leaders in the field of CSR towards the standard		Being able to get started without an external consultant	The content quality of the standard	The international acceptance of the standard
The support from stakeholders beside customers (NGOs, suppliers)		Suitability for newcomers in the area of CSR	Fully covering all aspects of CSR (People, Planet, Profit)	Providing support (such as providing information, advice, and management tools) by the organization that had developed the standard
The recommendation from consultancies		Name recognition of the standard	Promoting one definition for CSR by the standard	The involvement of stakeholders in the development process of the standard (how the standard was established)

(continued)

Table 1 (continued)

External forces/ requirements of the market	Organizational characteristics	Intangible characteristics of the standard	Tangible characteristics of the standard	Characteristics/ support of the standards organization
		increasing the legitimacy of an organization by means of the standard	Substitution ability of the standard (Note: Replacing other or already implemented standards	The legitimacy of the organization that has developed the standard
			improving the actual CSR performance of an organization	The expertise of auditors in auditing a CSR standard

In the second stage, the data gathered from the interviews was used for the design of a questionnaire aimed to use for quantitative purposes in order to verify the importance of the identified determinants for the adoption of a CSR standard on a larger scale (Taylor 2005; Creswell 2003), following a deductive research approach (Boeije 2005). The 31 determinants had to be scored by respondents on a 7-point Likert scale, ranging from 1 ('absolutely unimportant') to 7 ('absolutely important') by respondents. In order to obtain access to a group of corporate CSR professionals, we chose to tap the online business network LinkedIn. There is a specific Dutch LinkedIn group for CSR and sustainability professionals which has more than 15,000+ members and belongs to the top-20 groups in terms of members in the Netherlands. An invitation to participate in the survey was placed in this group. The response rate (as defined by the number of respondents divided by the total number of members) was not expected to be high since participation took place by self-selection (Wimmer and Dominick 2006), since this group is frequently used for research purposes by other researchers and students, and since many members of LinkedIn groups are passive members, merely 'lurking' information rather than actively contributing to groups. Still, we decided to pursue this option for collecting data as it is the largest repository of engaged CSR professionals and well accessible. Ultimately, in 2 weeks' time the number of respondents was 72 with all of the respondents having completed the full questionnaire.

The results of the quantitative part of our research are provided in Table 2. As the data show, only four out of the 31 determinants identified received an average score of 6 or more by the respondents. These are *Customer demand*, *Attitude of management*, *Content quality of the standard*, and *Actual improvement of CSR performance*. As Table 2 shows, the standard deviation of three out of these four determinants is below 1.000, suggesting a relative high degree of consensus among respondents. The categories *Intangible characteristics of the standard* and *Characteristics of the standards organization* did not contain determinants that received an average score of 6 or higher. The determinants that were perceived to be the least important factors influencing the adoption of ISO 26000 were *Recommendation from consultancies*

Table 2 Overview of determinants of the adoption of ISO 26000 (quantitative stage)

	N	Mean	Std. deviation	Variance
Determinant category: External market forces				
Customers	72	6.25	0.765	0.585
Early majority	72	5.03	1.048	1.098
Attitude of CSR leaders	72	5.28	1.024	1.049
Support stakeholders	72	5.44	1.352	1.828
Alternatives	72	5.00	1.444	2.085
Advice consultants	72	4.67	1.233	1.521
Determinant category: Organizational characteristics				
Attitude management	72	6.19	0.744	0.553
Experience with standards	72	5.31	1.030	1.060
Determinant category: Intangible standard characteristics				
Accessibility	72	5.92	0.900	0.810
Providing structure	72	5.19	1.229	1.511
Materiality to SMEs	72	5.53	0.964	0.929
Need for consultant	72	4.92	1.286	1.655
Suitability for newcomers	72	5.00	1.210	1.465
Name recognition	72	5.86	1.166	1.361
Increasing legitimacy	72	5.61	1.240	1.537
Determinant category: Tangible standard characteristics				
Costs	72	5.08	1.242	1.542
Certificate	72	5.31	1.607	2.581
Complementarity	72	5.53	1.100	1.210
Content quality	72	6.19	0.744	0.553
Covering PPP	72	5.86	1.190	1.417
Definition for CSR	72	5.36	1.541	2.375
Substitution ability	72	4.69	1.339	1.793
Improving actual CSR	72	6.08	1.071	1.148
Determinant category: Characteristics of standard organization				
Market position	72	4.92	1.330	1.768
Improvement of standard	72	5.33	1.088	1.183
Supplements	72	4.92	1.219	1.486
International acceptance	72	5.75	1.308	1.711
Providing support	72	5.17	1.374	1.887
Involvement stakeholders	72	5.75	1.286	1.655
Legitimacy of standard org	72	5.81	1.134	1.286
Expertise of auditors	72	5.67	1.278	1.634

(4.67), *Being able to get started without an external consultant* (4.92), *Substitution ability of the standard* (4.69), *Market position of the organization that developed the standard* (4.92), and *Provision of industry and other supplements by the organization that developed the standard or by industry associations* (4.92) Averaging the determinant scores per category shows that with a score of 5.75, *Organizational characteristics* is the category of determinants that scores highest (this category only contains two determinants, however), while the category *External market forces* scores lowest.

4 Discussion and Conclusion

Having empirically identified the most and least important factors determining the adoption of ISO 26000, we reflect on our findings from a theoretical point of view, confronting our data with several strands of literature. In doing so, we hope to contribute to a more thorough theoretical understanding of the phenomenon at hand.

The findings from our empirical data show that the requirements of customers were viewed of great importance for the adoption of a CSR standard, whereas the requirements of other stakeholders such as NGOs and suppliers were viewed as less important. From the perspective of institutional theory on isomorphism this implies that coercive market forces (DiMaggio and Powell 1983) are perceived stronger and therefore are more important for the adoption of a CSR standard by organizations than other determinants. The findings hence confirm Delmas' (2003) view that the actions and in particular the procurement activities of other firms play an important role in the adoption process since larger firms have the option to require their suppliers to adopt a certain standard. Next to the government and (large) companies with a clear CSR commitment, not many other market actors have defined CSR requirements for suppliers yet in the Netherlands. However, the analysis showed that these requirements were expected to be set by a larger number of companies in the near future as CSR was expected to keep picking up pace. Also, the support of the standard by other stakeholders such as NGOs appeared to play a role, while recommendations by consultancies were not expected to influence the adoption of ISO 26000 to an equally large extent. An explanation for this could be that consultancies are not powerful enough to function as a professional network and to generate a homogenous set of organizational norms such as a universal standard (cf. Delmas and Montes-Sancho 2011). In addition, consultants may have been viewed to have all too obvious commercial interest to be seen as powerful professional networks. The choice of an early majority for and the attitude of (for-profit) organizations that are seen as leaders in CSR towards the standard appeared to also play a role, indicating a certain degree of mimetic isomorphism to emerge among companies (DiMaggio and Powell 1983). This theory suggests that companies will imitate the structure and behavior of other organizations that are perceived to be more successful or have a high legitimacy, such as CSR leaders (cf. Suchman 1995).

The results showed that the accessibility of a CSR standard was a relatively important determinant for the adoption of ISO 26000 (cf. Rogers 1983, 2003). The provision of a structure by means of the standard, the content quality of the standard, and the actual improvement of CSR performance were viewed as important by the interviewed experts, which implies that a relative advantage might be perceived through these factors by the companies. Also, the implementation and certification costs can contribute to a perceived relative (dis)advantage. However, costs were generally viewed as less important from an adoption perspective, implying that the perceived benefits outweigh the perceived costs. Compatibility with existing practices and attitudes is reflected in the experience of organizations in working with standards and in the complementarity of the standard with other standards. The results show that both factors were seen as clear advantage though not as main determinants for adoption. The observability relates to the possibility of obtaining a certificate as well as to the name recognition of the standard. The importance of an independent external certification was assessed differently, but making a company's CSR performance visible either by means of a certificate or a self-declaration was generally seen as important, whereas the name recognition of the standard itself was less important. Diffusion of innovation theory (Rogers 1983, 2003) states that the different factors have to be perceived positively by an individual in order to allow adoption. Overall, the analysis confirmed Rogers' view that for the adoption of a CSR standard as an innovation, a company's management has to be persuaded of the different characteristics of a standard to perceive a clear benefit and to eventually promote adoption.

Economics of standards theory describes how standard wars take place and how an industry standard can be set by a firm. Shapiro and Varian (1999) identified several key characteristics of firms to outperform competition and to win the battle for the standard, such as control over an installed customer base, the ability to innovate, first-mover advantages, strengths in complements, and reputation and brand name. The market position of a standards organization, including the control over an installed customer base as well as the brand/company name of the standards organization, was assessed in the analysis to be of importance for the successful adoption of a standard. The control over an installed customer base can lead to a (more or less) direct adoption by that specific group, whereas the recognition of the brand or company name can lead to the adoption by companies outside the customer base. Reputation, in this case the legitimacy of the standards organization instead of the organization that is adopting the standard, was considered to be less important for the adoption. Also the continuous improvement of the standard, which demonstrates the ability to innovate of the organization behind a standard, was regarded as relevant but was not seen as a main driver for adoption. The ability to innovate might become more important at a later stage of adoption. Gaining a first-mover advantage in the market through an early majority which adopts a standard was regarded as important as well, but it proved difficult to assess in terms of actual influence. The fact that ISO 26000 and other CSR standards which have emerged in the Netherlands have some characteristics in common which makes switching easier, and therefore lowers a first-mover advantage (Moratis and Widjaja 2014). The strength in complements is reflected in the possible provision of supplements for specific industries and sectors, which was regarded as supporting the promotion of the standard. However, these sector supplements are yet to be developed by NEN or ISO.

Moreover, the degree of product imitation and the availability of complementary products are mentioned as key factors for successful adoption in this strand of theory (cf. Hill 1997). The analysis showed that the existence of alternatives for demonstrating CSR performance was acknowledged as well as the possibility to imitate ISO 26000 or develop a sector-specific or a version specifically suited for small and medium-sized enterprises that could be certified. However, few alternatives were seen as valuable and therefore as serious substitutes. Rather than substitutes these alternatives can be seen as complementary products for the assistance with implementation (e.g., a CSR performance monitoring tool). As the second to last factor, the expertise of auditors, which is very specific to certification standards, was not referred to in the theories, but the analysis showed that companies value it as an important factor for adoption. Companies assume that auditors are sufficiently capable for auditing a CSR standard. At the same time, some companies already appeared to have experienced auditors having difficulties with CSR as a new field for auditing or certification. This could have a negative influence on the demand for external auditing and certification. The international acceptance of a standard is actually derived from several factors: on the one hand from external market forces and from the standards organization and the standard itself on the other. International acceptance is generally expected to enhance adoption, but it seems to be far more important to larger, internationally operating organizations than to SMEs. Still, for dominantly nationally active SMEs, this may be relevant as they also supply to those companies.

A conclusion from this article is that institutional theory, emphasizing legitimacy-granting activities by firms, probably offers the best explanation in the context of the adoption of ISO 26000. However, it should be noted that other theoretical perspectives such as diffusion of innovations (Rogers 1983, 2003) and economics of standards theory (Shapiro and Varian 1999) on standards adoption provide useful insights as well and each of them highlights certain aspects of CSR standards adoption.

This article therefore suggests that in order to develop a full understanding of comprehensive CSR standards adoption dynamics, recognizing the complementarity of various theoretical perspectives is therefore essential. This conclusion could lead scholars in this field to combine the different perspectives into a more encompassing theory of CSR standards adoption. Such a framework could also speculate on related aspects, including different stages of adoption (cf. Tolbert and Zucker 1996; Delmas and Montes-Sancho 2011) and different types of comprehensive CSR standards, recognizing the difference between certifiable and non-certifiable modalities. These subjects should in our opinion receive further empirical attention as well. ISO 26000 may prove an interesting case in point in this respect as it provides organizations with a comprehensive (rather than an issue-oriented) CSR standard and as it has engendered a debate about the certification of standards. What makes ISO 26000 an even more attractive research subject is that recent developments in the domain of standardizing CSR include the development of certifiable alternatives for ISO 26000, both initiated from independent certification organizations as well as various national standards bodies that operate within the ISO governance system (Moratis 2016). This means that organizations that want to adopt a comprehensive CSR standard may have the choice between different and in effect competing models of CSR standardization.

Our empirical data may support firms in the process of selecting the most appropriate among competing CSR standards. The results from our research may also offer some suggestions for standards developing organizations. The insights we provide may help them develop CSR standards with beneficial characteristics (or revise existing standards) from the perspective of spurring the adoption of these standards among firms. Finally, this article may be relevant for policy makers wanting to encourage responsible business behavior. Policy on CSR may benefit from the empirical insights in the sense that they identify several 'buttons' for spurring the adoption of comprehensive CSR standards, which may make it easier for firms to start or accelerate their CSR initiatives.

In conclusion, as CSR standards may play an important role in developing the roles and responsibilities of business in society, we hope this article will prove to be an impetus to pursue both the development of theory and conducting empirical research on the adoption of these standards.

References

Bansal, P., & Hunter, T. (2003). Strategic explanations for the early adoption of ISO 14001. *Journal of Business Ethics, 46*(3), 289–299.

Boeije, H. (2005). *Analyseren in kwalitatief onderzoek: Denken en doen*. Amsterdam: Boom Onderwijs.

Brunsson, N., & Jacobsson, B. (2000). The contemporary expansion of standardization. In N. Brunsson & B. Jacobsson (Eds.), *A world of standards* (pp. 1–17). Oxford: Oxford University Press.

Castka, P., & Balzarova, M. (2008a). Adoption of social responsibility through the expansion of existing management systems. *Industrial Management and Data Systems, 108*(3), 297–309.

Castka, P., & Balzarova, M. (2008b). Social responsibility standardization: Guidance or reinforcement through certification? *Human Systems Management, 27*(3), 231–242.

Castka, P., & Balzarova, M. (2008c). ISO 26000 and supply chains – On the diffusion of the social responsibility standard. *International Journal of Production Economics, 111*(2), 274–286.

Christmann, P., & Taylor, G. (2006). Firm self-regulation through international certifiable standards: Determinants of symbolic versus substantive implementation. *Journal of International Business Studies, 37*(6), 863–878.

Corbett, C., & Kirsch, D. (2001). International diffusion of ISO 14001 certification. *Production and Operations Management, 10*(3), 327–342.

Creswell, J. (2003). *Research design: Qualitative, quantitative, and mixed method approaches* (2nd ed.). Thousand Oaks: Sage Publications.

Curkovic, S., Stroufe, R., & Melnyk, S. (2005). Identifying the factors which affect the decision to attain ISO 14000., *30*(8), 1387–1407.

Del Mar Alonso-Almeida, M., & Rodriguez-Anton, J. (2011). Organisational behaviour and strategies in the adoption of certified management systems: An analysis of the Spanish hotel industry. *Journal of Cleaner Production, 19(13*, 1455–1463.

Delchet-Crochet, K., & Vo, L. (2013). Classification of CSR standards in the light of ISO 26000. *Society and Business Review, 8*(2), 134–144.

Delmas, M. (2002). The diffusion of environmental management standards in Europe and the United States: An institutional perspective. *Policy Sciences, 35*(1), 91–119.

Delmas, M. (2003). *An institutional perspective on the adoption of international management standards*. Santa Barbara: Institute for Social Behavioral and Economic Research Publications.

Delmas, M., & Montes-Sancho, M. (2011). An institutional perspective on the diffusion of international management system standards: The case of the environmental management standard ISO 14001. *Business Ethics Quarterly, 21*(1), 103–132.
DiMaggio, P., & Powell, W. (1983). The iron cage revisited: Institutional isomorphism and collective rationality in organizational fields. *American Sociological Review, 48*(2), 147–160.
Drobny, N. (1997). Environmental management for the 21st century. In T. Tibor & I. Feldman (Eds.), *Implementing ISO 14000* (pp. 1–14). Chicago: Irwin.
Guler, I., Guillien, M., & Macpherson, J. (2002). Global competition, institutions, and the diffusion of organizational practices: The international spread of ISO 9000 quality certificates. *Administrative Science Quarterly, 47*(2), 207–232.
Hahn, R. (2012). Standardizing social responsibility? New perspectives on guidance documents and management system standards for sustainable development. *IEEE Transactions in Engineering Management, 59*(4), 717–727.
Hill, C. (1997). Establishing a standard: Competitive strategy and technological standards in winner-take-all industries. *Academy of Management Executive, 11*(2), 7–25.
ISO. (2010). *ISO 26000 – Guidance on social responsibility*. Geneva: ISO.
Leipziger, D. (2010). *The corporate responsibility code book*. Sheffield: Greenleaf Publishing.
Levis, J. (2006). Adoption of corporate social responsibility codes by multinational companies. *Journal of Asian Economics, 17*(1), 50–55.
McWilliams, A., Siegel, D., & Wright, P. (2006). Corporate social responsibility: Strategic implications. *Journal of Management Studies, 43*(1), 1–18.
Mena, S., & Palazzo, G. (2012). Input and output legitimacy of multi-stakeholder initiatives. *Business Ethics Quarterly, 22*(3), 527–556.
Miles, M., & Munilla, L. (2004). The potential impact of social accountability certification on marketing: A short note. *Journal of Business Ethics, 50*(1), 1–11.
Miles, M., Munilla, L., & McClurg, T. (1999). The impact of ISO 14000 environmental management standards on small and medium-sized enterprises. *Journal of Quality Management, 4*(1), 111–122.
Moratis, L. (2016). Signalling strategies for ISO 26000: A firm-level approach. *International Journal of Operations &Productions Management, 36*(5), 512–531.
Moratis, L., & Widjaja, T. (2014). Determinants for the adoption of ISO 26000: Exploring the case of ISO 26000 and the CSR Performance Ladder in the Netherlands. *Social Responsibility Journal, 10*(3), 516–536.
Mueckenberger, U., & Jastram, S. (2010). Transnational norm-building networks and the legitimacy of corporate social responsibility standards. *Journal of Business Ethics, 97*(2), 223–239.
Mueller, M., Gomes dos Santos, V., & Seuring, S. (2009). The contribution of environmental and social standards towards ensuring legitimacy in supply chain governance. *Journal of Business Ethics, 89*(4), 509–523.
Neumayer, E., & Perkins, R. (2005). Uneven geographies of organisational practice: Explaining the cross-national transfer and diffusion of ISO 9000. *Economic Geography, 81*(3), 237–259.
Nikolaeva, R., & Bicho, M. (2011). The role of institutional and reputational factors in the voluntary adoption of corporate social responsibility reporting standards. *Journal of the Academy of Marketing Science, 39*(1), 136–157.
Poksinska, B., Dahlgaard, J., & Eklund, J. (2002). Implementing ISO 14000 in Sweden: Motives, benefits, and comparison with ISO 9000. *International Journal of Quality & Reliability Management, 20*(5), 585–606.
Potoski, M., & Prakash, A. (2004). Regulatory convergence in nongovernmental regimes? Cross-national adoption of ISO 14001 certifications. *The Journal of Politics, 66*(3), 885–905.
Quazi, H., Khoo, Y., Tan, C., & Wong, P. (2001). Motivation for ISO 14000 certification: Development of a predictive model. *The International Journal of Management Science, 29* (6), 525–542.
Rogers, E. (1983). *Diffusion of innovations*. New York: Free Press.
Rogers, E. (2003). *Diffusion of innovations* (4th ed.). New York: Free Press.

Shapiro, C., & Varian, H. (1999). The art of standard wars. *California Management Review, 41*(2), 8–32.
Suchman, M. (1995). Managing legitimacy: Strategic and institutional approaches. *Academy of Management Review, 20*(3), 571–610.
Taylor, G. (2005). *Integrating quantitative and qualitative methods in research* (2nd ed.). Lanham: University Press of America.
Tolbert, P., & Zucker, L. (1996). The institutionalization of institutional theory? In S. Clegg, C. Hardy, & W. Nord (Eds.), *Handbook of organization studies* (pp. 175–190). London: Sage.
Wimmer, R., & Dominick, J. (2006). *Mass media research: An introduction*. Belmont: Thomson Wadsworth.
Zadek, S. (1998). Balancing performance, ethics, and accountability. *Journal of Business Ethics, 17* (13), 1421–1441.

Lars Moratis, Ph.D., is Professor of Corporate Social Responsibility and Sustainable Business at Antwerp Management School and NHTV Breda University of Applied Sciences. He holds a Ph.D. in management science and has worked in the field of CSR in both the academic and the business context. His research interests lie in the fields of sustainable business models, CSR strategy development and implementation, the credibility of corporate CSR claims, the adoption of CSR and sustainability standards, critical approaches towards CSR, the psychology of sustainability, and responsible management education. He has authored and edited several books, chapters, academic and practitioner-oriented articles, and opinion pieces on these topics.

Alice Tatang Widjaja, studied business administration at the University of Amsterdam where she graduated on a thesis about ISO 26000. She is currently working as a product owner at Evellia in France.

ISO 26000: Is It a Case of Implementation or Simple Reporting? An Exploration of CSR Standards in Companies in Romania

Claudiu Bocean and Catalina Sitnikov

1 Introduction

The increased awareness of human rights and labor rights, environmental pollution, driven by the rapid expansion of communication technologies has spurred an increased interest in the responsible corporate behavior, both from the point of view of companies and governments (Kaphengst et al. 2013). If the first requirement of organizations is to respect the laws of the country where they operate, in addition to this requirement companies are requested to solve problems related to the expectations of stakeholders and society as a whole.

Social responsibility requires taking into consideration the expectations of all stakeholders as well as of the management of all economic, social and environmental issues which arise from relationships with stakeholders in their areas of influence: the workplace, the market, the adding value chain, community and public policies. Social responsibility is the solution to compatibilize the company's profit objectives with its social objectives. Attention to social problems may help maintain or improve organizational ethics, increasing the company's market value and appreciation of local community (Heal 2004).

Corporate Social Responsibility (CSR) is one of the latest areas that are targeted by corporate behavior regulatory processes. In the center of this regulatory process is the development of CSR standards which seek to harmonize ways for implementation and report (Gilbert et al. 2011; Waddock 2008).

Both at national and international level, there are CSR initiatives that have explicit support from the government. In addition, a variety of private international organizations were involved in the launching of CSR standards (Gilbert et al. 2011; Waddock 2008). Initiatives in CSR standardization, such as SA 8000, GRI and ISO

C. Bocean (✉) · C. Sitnikov
Faculty of Economics and Business Administration, University of Craiova, Craiova, Romania

S. O. Idowu et al. (eds.), *ISO 26000 - A Standardized View on Corporate Social Responsibility*, CSR, Sustainability, Ethics & Governance,
https://doi.org/10.1007/978-3-319-92651-3_5

26000, PRI are considered promising approaches regarding social issues and the environment. They are useful tools in the standardization process of CSR implementation and reporting (Tschopp and Nastanski 2014). However most of these CSR initiatives have failed to establish effective mechanisms which diminished their credibility and acceptance (Kaphengst et al. 2013). Functional inefficiency and voluntary nature generate a high danger of formal compliance, which involves the communication of increased social and environmental performance on the background of failure to meet standards. Avetisyan and Ferrary (2012) consider that extending the use of different standards in CSR at different levels creates a number of problems: the multitude of specific requirements of CSR standards that may conflict, the disregard of many issues addressed by various standards, the necessity to ensure interoperability of different standards.

The instruments and initiatives from the social responsibility area that can be used by organizations are conventions, declarations, principles and guidelines issued by international bodies, codes of conduct, management standards, tools and reporting standards, tools and national standards created by the national legislation. At the international level, there are four CSR initiatives that are recognized by most governments. They are: The ILO Declaration on Fundamental Principles and Rights at Work, the OECD Guidelines for Multinational Enterprises (MNEs), the UN Global Compact (UNGC) and the UN Guiding Principles on Business and Human Rights (UNGP). They offer detailed recommendations on principles and approaches to CSR (Kaphengst et al. 2013). Codes of conduct can be used as general milestones in the organization, guidelines which align present business practice to the international practice which should be considered especially when the national legislation in a particular field is not comprehensive enough. Management standards (ISO 14001, SA 8000, Account Ability AA1000 Series, ISO 26000, etc.) are the documents that set the acceptable levels of performance in social responsibility and may be voluntarily adopted by enterprises (Szatmari et al. 2012). Based on these standards, certifications and authorizations from third parties can be obtained (such as AA1000, SA 8000 or ISO 14001) this offering credibility to enterprises towards stakeholders. Using the tools and reporting standards, organizations can communicate to stakeholders their performance in the field of social responsibility and, at the same time, can yearly measure the trend of these performances. Steps towards social responsibility are made, firstly, by conforming to the law, which makes useful and necessary the knowledge of legislative framework of the country in which the organization operates.

The chapter contains five sections. The first section provides an introduction to the research issue, while the second section describes the research methodology. The third section focuses on the analysis of organizations' social responsibility in Romania. Section 4 is intended to describe CSR communication policies and ensuring transparency of organizations in Romania regarding CSR actions and the used social responsibility tools. In the fourth section we have identified the six main instruments and CSR standards. Section 4 is devoted to the quantitative analysis of the use of instruments and CSR standards within the Romanian organizations. Section 5 provides conclusions.

2 Literature Review

In Romania, the concept of CSR appeared in the 1990s with the foundation of NGOs that had in particular a humanitarian purpose. This phenomenon was followed in 2000 by the Romanian restructuring processes as a result of the preparation for Romania's accession to the European Union. These restructuring processes have involved the replacement of production technologies with more "environmentally-friendly" ones, and massive layoffs which generated social needs in relation to those workers which could not find a job.

The implementation of CSR practices by the private sector has become stronger after the integration into the European Union and was dominated by multinationals who have transferred practices and organizational culture from headquarters to local level and later by the opportunity to access European funds. In addition to knowledge and best practice transfer, other companies and a "responsible industry" have developed since the European integration—companies whose business objective itself is "sustainability". These companies are engaged in business activities which, by their nature, bring environmental and/or communities' benefits. In this case, their "responsibility" indirectly results from their goods and services they produce and provide (Anca et al. 2011).

As a result of increased competition in the Romanian business environment as well as of the extension of the influence of the multinational companies, the organizations which carry out their activities in Romania are beginning to realize the importance of CSR activities (especially when they are integrated into the organizational strategy) in the process of improving their image and increasing the reputation of the organization. Thus, more and more organizations in Romania have set up departments to implement the CSR activities. In other situations, although specific departments haven't been established, the CSR activities are implemented by other departments, such as communication, marketing, public relations. Following the research carried, Obrad et al. (2011) found that within the Romanian organizations CSR activities are developed in most cases by a special compartment within a specific department, usually the PR or the Communication department. Only in isolated cases, particularly in multinational companies, this activity takes place within a specific CSR department. In most cases, the CSR activity is coordinated by a department manager, and only in special cases, this function is taken over by the General Director or by one of the board members. According to the study conducted by Ernst and Young (2016) in 2015 68% of organizations in Romania place the budget line for CSR in the department of PR and Marketing, 12% in HR departments, and only 13% have a separate budget for sustainable development or CSR. Concerning the number of people involved in the CSR team in 2015 there were on average 2.9 per company (full time employees), which indicates a decrease of 17% compared to 2014 (on average 3.5 full time employees per company).

Another phenomenon that characterizes the Romanian company is setting up the charitable foundations of large organizations that are wholly or partly funded by the organization and aimed at solving social or environmental issues that have great importance for the local community. These foundations take over the charities within the organization, at the level of organizations' CSR departments remaining strategy development and implementation of specific CSR programs (Obrad et al. 2011).

Responsible corporate behavior in Romania was initially associated with commercial considerations, image and reputation, not with sustainable development and meeting the needs of stakeholders. In recent years it was noticed a greater emphasis placed on the second approach, the CSR programs having a more strategic character aiming a long term vision.

Social responsibility plays an important role in strengthening the strategic position of the organization, in policies' differentiation process. A major challenge of large organizations relates to the development of a credible and reliable social responsibility strategy, which is based on clear and ethical principles and values. The promoters of these policies are generally multinational companies which implement their centrally-set strategies at the CSR national level. Social responsibility programs implemented by these multinational companies in Romania are mainly influenced by corporate culture and are rarely adapted to local needs. Usually these social responsibility policies implemented by large multinational corporations take the form of social assistance programs, social programs funding the humanitarian foundations, donations and educational support for those in need (European Commission 2007). Community projects or charity campaigns are most attractive to companies because these types of initiatives create awareness among community members. According to a study conducted by a consulting company (The Azores 2016) in Romania, companies give highest interest in the area of CSR community involvement (22% of companies evaluated). At the opposite end, companies give the slightest interest to the value chain, human rights and anti-corruption policy.

Crișan-Mitra and Borza (2015), in a study conducted on a sample of 87 respondents, selected among top managers, HR managers and heads of CSR departments within large companies operating in Romania, noted that managers are aware that CSR practices enable organizations to improve their image and reputation, causing not only positive reactions within the local community and employees, but also by strengthening the company's values, improving culture and organizational climate. According to the study conducted by Ernst and Young (2016) in 2015, 88% of companies surveyed say they were involved in projects of corporate social responsibility in 2015 versus 69% in 2014. This result marks a greater interest from companies to integrate CSR activities in their business strategy and minimize the effects of the economic crisis. The economic crisis has had a major impact on the willingness of organizations to use financial and human resources in areas not directly related to their core business. This has led to reluctance to invest in complex and strategic CSR activities with long-term impact.

Obrad et al. (2011) observed that there is no complete understanding of the concept of CSR among Romanian organizations because they do not have a culture that could create value and promote the principles of social responsibility in business.

Ernst and Young (2015) pointed out, after conducting a SWOT analysis, the weaknesses of CSR in Romania: the CSR concept is little known in the Romanian society, the absence of studies on awareness and application of CSR in Romania, some SMEs are not yet convinced of the relevance of CSR initiatives, lack of visibility of CSR efforts into stakeholders' groups. In their turn, Obrad et al. (2011) identified a number of barriers in the development of CSR in Romania: no qualified human resources in this area; the meaning of CSR is not clearly understood by the organizations; there is not pressure from community on organizations concerning solving social and environmental issues; companies are not involved in CSR, since competitors are not involved in such activities; insufficient budgets allocated to these activities. Starting from these barriers and Ernst & Young's allegations, we highlighted major weaknesses in the area of CSR in Romania: lack of specialized human resources in the field of CSR; lack of a legal framework to encourage and contribute to the development of CSR; CSR strategy is not integrated in the basic strategy of the organization; poor reporting and monitoring of CSR activities; the Romanian CSR is an imitation without too complex adaptation of foreign practices; CSR is mostly used as a PR tool as there is no long-term thinking.

In Romania, the development of social responsibility area is still at an early stage. While multinationals and large Romanian companies have gradually created a culture of accountability, in many cases, social responsibility is still associated with philanthropy, sponsorship and public relations activities. A small number of companies have adopted a strategic approach, integrating social responsibility into their core practices, influencing the decisions and activities of the company (Anca et al. 2011). The Azores (2016) finds that the best performing companies in Romania on social responsibility are, according to their index, multinational companies that have published a CSR report locally. Although the number of reports is reduced, most are developed in accordance with international reporting standards, the most common being the Global Reporting Initiative. Stakeholder engagement is a very important step in building the CSR strategy, as companies need to fully understand the potential risks to prevent them. However, organizations from Romania do not involve stakeholders enough in the decision making on investment in CSR initiatives and do not perform detailed analyzes to determine their needs.

The ways in which organizations address CSR depend on the strategic vision and key values of the culture of the organization. The current, rather restrictive, economic conditions determine organizations to reduce their investments, often to the detriment of social and environmental causes. In Romania, due to the lack of a clear legal framework, multinational companies face problems in transposing CSR policies practiced at central level to national or local communities. But according to a directive of the European Commission, since 2017, large companies which are public- interest entities and which at the balance sheet date exceed an average number of 500 employees during the financial year, must include in the management report a non-financial statement (containing information concerning at least the environmental, social and personal issues, respecting human rights and the fight against corruption and bribery). The structure of this statement must comprise a brief description of the company's business model; a description of the policies adopted by the company in connection with the non-financial issues; the results of those policies; the main risks related to these issues arising from company operations; non-financial key performance indicators (European Commission 2014).

If the company does not implement policies regarding one or more of these issues, non-financial statement provides a clear and reasoned explanation regarding this option. Member States may require non-financial statement information to be verified by an independent service provider (the European Commission 2014).

According to the Association for Community Relations, unlike the initial proposal, the text under scrutiny maintains a single criterion of applicability of the law, that of the size, established in terms of number of employees. Thus, transparency provisions will be applied to companies and public interest entities with more than 500 employees, i.e. at least 6000 companies, according to estimates by the European Council (Euractiv.ro 2014). In Romania, according to this criterion, it is estimated by the Association for Community Relations that 720 companies will be targeted based on the data obtained from the Trade Register (Euractiv.ro 2014).

The European Directive on non-financial reporting will have a positive impact on organizations in Romania, both on the multinational and national companies, giving them a clear and coherent framework for the reporting of CSR.

In its annual survey entitled "Trends and Realities CSR in Romania" Ernst and Young (2016) revealed that in 2015 over half of companies (51%) say they talked internally about the European directive on the non-financial reporting and its impact on the company. 28% of respondents came to the conclusion, as a result of discussions, that they will have to report under this Directive. However, 33% of respondents say they haven't discussed internally this issue, while 16% do not know if such consultations have taken place within the company.

In order to be effective, CSR strategy needs an effective communication policy. Only in this way it can improve the image and increase the organization's reputation, it can improve the feedback received from stakeholders and CSR may acquire a strategic character. CSR communication and transparency are default conditions to get expected results. To gain greater credibility, usually CSR reports take into account key international indicators and international standards such as GRI G4, AA 1000, SA 8000, ISO 26000, etc. The ways in which organizations address CSR depend on the strategic vision and key values of the culture of the organization. The rather restrictive current economic conditions determine organizations to reduce their investments, often to the detriment of social and environmental causes. In Romania, due to the lack of a clear legal framework, multinational companies face problems in transposing CSR policies practiced at the central level to national or local communities. But according to a directive of the European Commission, since 2017, large companies which are public interest entities and which at the balance sheet date exceed an average number of 500 employees during the financial year, must include in the management report a non-financial statement (containing information concerning at least the environmental, social and personal issues, respecting human rights and the fight against corruption and bribery). The structure of this statement must comprise a brief description of the company's business model; a description of the policies adopted by the company in connection with the non-financial issues; the results of those policies; the main risks related to these issues arising from company operations; non-financial key performance indicators (European Commission 2014).

Concerning standardization in CSR area, CSR initiatives which have experienced the most internationally widespread and have been embraced also by the Romanian companies are: ISO 14001, OHSAS 18001, SA 8000 Series AccountAbility AA1000, ISO 26000 and GRI G4 (Iamandi and Munteanu 2011; Obrad et al. 2011; Filip et al. 2012; Szatmari et al. 2012).

ISO 14001 is part of the ISO 14000 series of standards covering a wide range of issues: practical guidelines for analysis, training, audit, certification in environmental management. This standard developed by the International Organization for Standardization (ISO) describes the basic requirements of an effective environmental management system that allows continuous improvement of environmental performance and compliance with legislation (ISO 2015).

OHSAS 18001 is an international standard on Occupational Health and Safety Management System, which offers the necessary measures to control those risks in health and safety and improving performance in this area.

Social Accountability 8000 Standard (SA 8000) was created with the intention to provide a standard based on UN Declaration of Human Rights, the rules and international law in the area of labor, in order to protect and increase the importance of (SAI 2014). SA 8000 application is verified through an evaluation process based on evidence.

AccountAbility AA1000 series of standards is based on principles that help organizations become more responsible, sustainable and conscious. These standards address issues related to governance, business models and organizational strategy, provide guidance on sustainability and stakeholder involvement. AA1000 standards are designed to provide an integrated thinking on sustainability, being a support in the process of sustainability reporting activities (AccountAbility 2008).

ISO 26000 provides guidance on social responsibility without providing specific requirements or require certain results. Organizations that implement ISO 26000 are given the opportunity to identify their own priorities on social responsibility and build a sustainable business model in the spirit of a continuous improvement process (Ecologia 2011). ISO 26000 is not a standard certification but it offers a guideline to social responsibility (ISO 2010). At the same time, it is not subject to a management system but it can be integrated into other systems. Considering the broad area of addresses issues, ISO 26000 is considered as the most inclusive and most well-known among all the instruments of CSR.

Global Reporting Initiative (GRI) is a non-governmental organization that develops guidelines on the sustainability reporting (economic, environmental, social and corporate governance) being the reference tool in the area (GRI 2013). The main aim of the GRI is the harmonization of how sustainability reporting is carried out, to ensure comparability and analysis of corporate sustainability performance (Berinde and Andreescu 2015).

In order to determine the implementation of CSR management standards in Romania Filip et al. (2012) have considered 40 responsible companies voluntarily registered on the Romanian specialized website dedicated to CSR (www.responsabilitatesociala.ro)

and have analyzed them according to six standards relevant to CSR: AccountAbility's AA1000, G3 GRI, ISO 14001, ISO 26000, SA 8000, UNCG. Of the 40 companies that claimed to be responsible on the website, 19 had not implemented any of the six instruments or standards, the others have implemented one or more tools or management standards (maximum five in the case of Orange Romania). The research was conducted on the basis of information found on companies' web sites. Given that most of the 40 companies were not listed on the Stock Exchange (in the case of companies such as Orange Romania, only the parent company is listed on the Stock Exchange) they were not obliged to publish information on their activities, CSR communication being made voluntarily and randomly. In this way, we can explain the large number of companies declaring themselves as responsible, but that did not implement any CSR tools or standards. Filip et al. (2012) research results show that the most used tool in the area of CSR in Romania was ISO 14001 (16 companies out of 40), followed by GRI G3 (10 companies out of 40) and UNCG (5 companies out of 40). The most active sectors in terms of application of CSR tools and standards were Oil & Gas, Communications, Food & Beverages (Filip et al. 2012). Also, Filip et al. (2012) show that a good financial position in the market does not necessarily imply the application of CSR tools and standards.

Crişan-Mitra and Borza (2015) state that, in Romania, in the perception of organizational managers, the implementation of CSR standards does not significantly influence the purchase decision of the customer, as customers in Romania focus mainly on price and features of products and services, being less sensitive to responsible behavior due to a lower degree of awareness concerning social and environmental issues. These authors believe that this phenomenon is specific to countries with low living standards. In this type of countries, consumers are not willing to pay higher prices to support CSR activities compared with those where the standard of living is higher.

3 Research Methodology

To identify the main CSR tools and standards within the Romanian organizations, and the ways of their use and communication in public space, we conduct a two-dimensional research. The first dimension is qualitative and it aims to analyze social responsibility of organizations in Romania and CSR communication policies regarding CSR actions and social responsibility tools used. The second dimension is quantitative and it aims to identify the instruments and standards of CSR used in the Romanian organizations and how they communicate CSR policies implemented and the instruments used.

For the quantitative analysis we selected the ten companies included in the BET index of the Bucharest Stock Exchange (BSE). BET is the reference index of the capital market in Romania. BET reflects the evolution of the most traded companies on the BSE regulated market. The main criterion for selection of the companies in this index is liquidity. The choice of these companies is justified by the fact that, since 2015, for inclusion in the index, criteria have been applied related to transparency, quality reporting and communication with stakeholders. This allows us to state that our research results have a high relevance for large organizations from Romania.

Regarding the communication of CSR policies we start (based on the results of previous research: Obrad et al. 2011; Filip et al. 2012; Crişan-Mitra and Borza 2015) from the hypothesis formulated as follows:

H1 The most important Romanian organizations, companies listed on the BSE (included in the main index—BET) have effective communication of implemented social responsibility policies and used instruments covering all areas of CSR.

Regarding the identification of CSR instruments and standards used in Romanian organizations we have formulated the following hypothesis (based on the results of previous research: Iamandi and Munteanu 2011; Obrad et al. 2011; Filip et al. 2012; Szatmari et al. 2012):

H2 The most important Romanian organizations (companies listed on BSE in the main index—BET) use at least one instrument or CSR standard in implementing the programs of social responsibility and non-financial reporting. An additional hypothesis aimed to investigate relationship between the corporate culture of the organization and the standards and tools used to implement and report on CSR.

To test the validity of the hypotheses, we conducted an exploratory study on the way in which companies communicate through corporate website involving in CSR and CSR standards used. Research on the companies' websites was systematic and heuristic.

With the adoption since 2017 of the provisions of EU Directives 2014/95/ EU and 2013/34/EU concerning non-financial disclosure of large companies there will be many companies that will have a transparent policy regarding the used CSR policies and instruments. In the future the research in this area will be expanded.

4 Analysis of the Used CSR Tools and Standards in Romanian Organizations

Our study focuses on the most transparent organizations operating in Romania (companies included in the BET index of the Bucharest Stock Exchange—BSE). BET is the reference index of the capital market in Romania. BET reflects the evolution of the most traded companies on the BSE regulated market, excluding financial investment companies. The main criterion for selection of the companies in this index is liquidity. Since 2015, for selection, there have been applied the criteria related to transparency and quality reporting of issuers and their communication with investors (BSE 2016).

To check the validity of the hypothesis H1 we performed an analysis on the websites of ten companies included in the BET. We conduct this research on websites taking into account the results of Ernst and Young (2016) research according to which the predominant tool used to communicate the involvement in CSR remains the company's website (66% of respondents argued this). Under hypothesis H1, we tried to identify how organizations communicate CSR activities and analyze how organizations provide this information (annual reports, sustainability reports, codes of ethics, best practices, etc.). Also we conducted an exploratory research on the communication of organizational culture. The results of research are summarized in Table 1.

Fondul Proprietatea (FP) is an investment fund created in 2005 by the Romanian government to provide compensation for people who were abusively dispossessed of their properties by the communist authorities during the nationalization of the Romanian economy in the late 1940s after the Second World War. Since September 2010, the Fund has been managed by Franklin Templeton Investment Management Ltd. United Kingdom, Bucharest Branch. The organization's web page does not contain information on organizational culture or social responsibility actions taken by the organization. It only contains compulsory information requested by BSE on reports to investors (usually financial information).

Banca Transilvania (BT) is a bank that is ranked on third position among banks in Romania, in terms of assets. Organizational culture is not well highlighted on the company's website: mission statement is made in the text of the bank presentation and references are made on the elements of organizational culture in different sections of website. Banca Transilvania does not have a section dedicated to CSR in the main web page, but in the news section and a blog of the organization which displays sponsorships and local community projects.

OMV Petrom (SNP) is the largest integrated oil and gas producer in Southeastern Europe, with activities in Upstream, Downstream Gas, Downstream Oil sectors. The entire activity of OMV Petrom is based on the principle of sustainability (individualized as the resourcefulness at company level) under which it is targeted profitable growth in a responsible manner. OMV Petrom puts particular emphasis on the idea of responsible and sustainable behavior in business, and constantly builds relationships based on trust and honesty with all stakeholders. Organizational culture is well highlighted on the website. Website displays vision and mission, values, code of conduct (acquired from OMV and which is based on UNGC principles). The sustainability section of the website is well organized and communicates all OMV Petrom initiatives in social responsibility area: education, environment, recognition and awards, local initiatives, business ethics, employee engagement, human rights.

BRD is the second bank in Romania in terms of assets, one of the major lenders of SMEs and one of the major players in the corporate banking market in Romania. BRD affirms its commitment to the communities where it operates, through actions in support of disadvantaged and social vulnerable children and young people and through reducing activity impact on the environment having a responsible management of resources such as energy, water and paper. Elements of organizational culture are well highlighted, excluding code of conduct which is not made public. On the website, there is a section describing social responsibility projects and

Table 1 Communication policy of CSR activities and organizational culture of organizations included in the BET index of BSE

BSE symbol	Organization	Vision, mission	Values	Norms/code of conduct	CSR or sustainability section	Donations, sponsorship section	Local community section	Other CSR actions communicated
FP	Fondul Proprietatea	–	–	–	–	–	–	–
TLV	Banca Transilvania S.A.	x	–	–	–	x	x	x
SNP	OMV Petrom S.A.	x	x	x	x	x	x	x
BRD	BRD—Groupe Societe Generale S.A.	x	x	–	x	x	x	x
SNG	S.N.G.N. Romgaz S.A.	x	x	–	x	x	x	x
EL	Societatea Energetica Electrica S.A.	x	x	x	x	x	–	x
TGN	S.N.T.G.N. Transgaz S. A.	x	–	–	x	x	–	x
TEL	C.N.T.E.E. Transelectrica	x	–	–	–	x	–	–
SNN	S.N. Nuclearelectrica S. A.	x	x	–	x	–	x	x
BVB	Bursa de Valori Bucu-resti SA	x	–	–	x	–	–	x

– denotes lack of communication elements on websites
x denotes the existence of communication elements on the websites

programs concerning community investment, employee volunteerism, responsible funding, and environmental policy.

Romgaz (SNG) is the largest producer and main supplier of natural gas in Romania. Romgaz mission contains only elements of sustainability regarding the relationship with the shareholders: continuous growth performance, competitiveness and value of company due to the potential of human resources and assets held predictable and profitable business practices and a rigorous risk management. Except for the conduct code, all the organizational culture elements are communicated. CSR section provides information on projects in various areas: environment, community, education, health, culture, sports, and a sponsorship guide.

Electrica (EL) is the leader in distribution and supply of electricity in Romania and one of the most important players in the energy service sector. All elements of organizational culture within Electrica are very well highlighted. There is issued a Code of Ethics and Professional Conduct. There is a mechanism for whistleblowers. CSR initiatives are communicated in a dedicated section of the website. Within it, sponsorships and donations are highlighted.

Transgaz (TGN) aims to fulfill the national strategy established for transport, international transit and dispatching natural gas. The responsibility policy aims to increase permanently the accountability of the company towards employees, shareholders, partners, community and environment as streamlining of CSR programs impact initiated for this purpose. CSR actions are summary displayed on the web page, insisting on sponsorship activity. The organizational culture is also communicated in a summary manner, the focus being on communicating their mission and vision.

Transelectrica (TEL) provides transport and system services for electricity in the national energy system. Both organizational culture and social responsibility are disclosed in a summary manner, there is only a section on sponsorships and donations.

Nuclearelectrica (SNN) aims to produce electricity, thermal and nuclear fuel. The communication of organizational culture is appropriate and presents the mission, vision and values of the company. CSR communication is quite brief missing sections devoted to sponsorship, yet including a section devoted to local community projects.

Bucharest Stock Exchange (BSE) is the stock exchange in Romania capital market. BSE provides funding opportunities for companies that want to attract investment capital market in Romania. CSR communication and organizational culture is quite brief, resuming at the transmission of mission, vision and overall CSR policies. In March 2015 the Bucharest Stock Exchange (BSE) has become partner of the United Nations initiative, Sustainable Stock Exchanges (EEA). The initiative is supported by the UN Principles for Responsible Investment, the United Nations Conference on Trade and Development, the United Nations Environment Programme Finance Initiative and the UN Global Compact (BSE 2016).

As a result of hypothesis H1 research, we have found that the most important Romanian organizations listed in BSE (included in the main index—BET) have a relatively efficient communication of implemented social responsibility policies and

Table 2 CSR management standards and tools used and communicated by organizations included in the BET index of BSE

BSE symbol	Organization	Share in BET index	ISO 14001	OHSAS 18001	SA 8000	AA 1000	ISO 26000	GRI G4
FP	Fondul Proprietatea	21.15	–	–	–	–	–	–
TLV	Banca Transilvania S.A.	19.98	–	–	–	–	–	–
SNP	OMV Petrom S.A.	14.87	x	x	–	–	–	x
BRD	BRD—Groupe Societe Generale S. A.	11.76	–	–	–	–	–	–
SNG	S.N.G.N. Romgaz S.A.	10.39	x	–	–	–	–	–
EL	Societatea Energetica Electrica S.A.	8.98	x	x	–	–	–	–
TGN	S.N.T.G.N. Transgaz S.A.	6.74	x	–	–	–	–	–
TEL	C.N.T.E.E. Transelectrica	4.14	x	x	–	–	–	–
SNN	S.N. Nuclearelectrica S. A.	1.14	x	x	–	–	–	–
BVB	Bursa de Valori Bucuresti SA	0.86	–	–	–	–	–	–

instruments used. The hypothesis was validated in part because there are few companies that have poor communication on CSR (TEL and BVB), and one of them completely ignores CSR (FP).

To check the validity of the hypothesis H2 we conducted an analysis of CSR tools and standards that companies included in the BET index of BSE used to address social responsibility and how they communicate CSR policies and programs. The result of research is summarized in Table 2. In the same hypothesis, we investigated whether a relationship between the corporate culture of the organization and the standards and tools used to implement and report on CSR can be established.

Fondul Proprietatea (FP) and Banca Transilvania (BT) do not communicate anything concerning the CSR instruments and standards used in their activities. OMV Petrom (SNP) is the company of the BET index of BSE using a wide range of instruments and standards of social responsibility: ISO 14001, OHSAS 18001, and GRI G4. Among the tools of social responsibility BRD uses, as a subsidiary of Societe Generale, the Equator principles to make analysis of all investments larger than US$10 million. None of the tools included in our analysis are used by BRD.

Among the CSR instruments and standards analyzed, Romgaz (SNG) uses only ISO 14001 indicating a sustainable policy in the area of environmental protection. Electrica (EL) uses an integrated management system, built by incorporating elements of integrated management system quality-environment-security implemented and certified in accordance with ISO 9001, ISO 14001, OHSAS 18001. Transgaz (TGN) uses only ISO 14001 among the CSR instruments and standards analyzed. As Electrica, Transelectrica (TEL) uses an integrated management system that includes ISO 9001, ISO 14001, OHSAS 18001. Also, Nuclearelectrica joins energy companies regarding the use of an integrated management system which includes the three standards ISO 9001, ISO 14001, OHSAS 18001. Although the Bucharest Stock Exchange (BVB) is part of the Sustainable Stock Exchanges initiative (EEA), this company is not communicating on CSR standards and tools, focusing on the principles of corporate governance that it proposes to all issuers.

As a result of the investigations, we have found out that companies listed in the main index of BSE do not use many CSR tools and standards of social responsibility. The most used are those that allow certification (ISO 14001 and OHSAS 18001). Although in turn they allow certification, SA 8000 and AA 1000 standards are not used by any of the ten companies analyzed. Another important aspect that we have found is the fact that none of the companies take into account ISO 26000. The only one of the ten companies which uses reporting tool for non-financial information (GRI G4) is OMV Petrom. We also found that the corporate culture does not dictate the implementation of CSR management standards. The results are consistent with the main findings from the literature (Iamandi and Munteanu 2011; Filip et al. 2012). The influence of organizational culture works only in terms of benchmarking within local subsidiaries of multinationals' CSR practices. Domestic companies are not very active in the area of the use of CSR tools and standards. Another variable affecting the use of CSR tools and standards is the industry sector. It is noted that the energy companies are more active in terms of implementing CSR standards and tools than those in the financial sector. This is due to the complex environmental problems facing that sector.

In conclusion, H2 hypothesis is partially validated. Of the ten companies analyzed, seven use at least one instrument or standard management of CSR (ISO 14001 is mostly used). A significant aspect is the use of ISO 26000 in the most important Romanian companies listed on BSE. None of the companies analyzed use ISO 26000 and any other instrument covering all the issues of social responsibility.

5 Conclusions

Corporate social responsibility (CSR) has become an important policy for each large organization that wants to maintain an adequate relationship with its stakeholders (customers, community, business partners, local authorities, etc.). Even if the undertaking of social responsibility programs is not compulsory, a socially or environmentally irresponsible behavior can affect not only the image of the organization, its perception in the market but also its strategic approach. The approach of social responsibility policies is influenced by the values, motivations, decisions of managers of the specific organizational culture and business models that are found within a national economy.

In the last decade, the Romanian organizations have made progress in adopting and implementing decisions and activities regarding social and environmental issues. Despite this growth, most organizations continue establishing their social responsibility programs only from sponsorships and random charitable projects. Although the role of sponsorships, donations and random initiatives is beneficial, the holistic approach of social and environmental issues requires a voluntary integration of social responsibility principles and practices in the core activity and in the relationships with stakeholders, a proactive implementation of such principles and values. Such behavior implies a strategic integration of social responsibility in organizations' practices. Strategic integration of social responsibility into core business makes responsible behavior to be less vulnerable to external factors (such as budget cuts in times of economic crisis), thus becoming an integral part of the business model.

In this chapter we conducted an analysis of the most important Romanian organizations listed at the BSE (included in the main index—BET). We concluded that companies listed in the main index of BSE do not use many CSR tools and standards, the most used being ISO 14001 and OHSAS 18001. Companies are not interested in the implementation of ISO 26000 because the action involves a large consumption of resources (financial, human and time) and the benefits are limited given the low level of awareness on social responsibility of the Romanian consumers. CSR communication is made by most companies but in many cases it is a formal activity, CSR activities not being integrated into the core business strategies.

The implementation of CSR tools and standards enhances the reputation and the image of the responsible organization and is a source of competitive advantage. This is one of the reasons for which organizations should adopt CSR tools and standards and communicate publicly on the use of these tools and standards. In this regard, one of the main requirements for companies acting on the Romanian business environment is to strengthen the implementation of CSR standards and communicate about them.

Studying the influence of organizational culture on the application of CSR tools and standards in Romania, we observe that multinational companies apply to their subsidiaries in Romania the same policies and standards as in their home countries. Domestic companies only sporadically implement CSR tools and standards having an imitation behavior. Also, it can be seen that companies from the energy sector are more inclined to use CSR tools and standards than the companies from the financial sector which are more focused on the relationships with shareholders than the ones with other stakeholders.

References

AccountAbility. (2008). *AA1000 AccountAbility principles standard*. Retrieved October 16, 2016, from http://www.accountability.org/standards/

Anca, C., Aston, J., Stanciu, E., & Rusu D. (2011). *Responsabilitate Socială Corporativă în România. Analiză situațională și recenzie a practicilor actuale* [Corporate Social Responsibility in Romania. Situational Analysis and Review of Current Practices]. Raport realizat în cadrul Proiectului POSDRU/64/3.3/S/41722 Întărirea capacității companiilor românești de dezvoltare a parteneriatelor sociale – RSC [Report within SOPHRD/64/3.3/S/41722 Project-Strengthening the Capacity of Romanian Companies to Develop Social Partnerships-CSR]. Retrieved September 29, 2016, from http://undp.ro/libraries/projects/CSR/deliverables/Analiza%20Situatiei%20RSC%20in%20Romania%20RO(1).pdf

Avetisyan, E., & Ferrary, M. (2012). *Dynamics of stakeholders' implications in the institutionalization of CSR field in France and in the United States*. Paper presented at Annual Academy of Management conference in Boston, 3–7 August 2012.

Berinde, M., & Andreescu, N. A. (2015). Reporting corporate social responsibility according to GRI standards. The Annals of the University of Oradea. *Economic Sciences, 24*, 17–23.

Bucharest Stock Exchange (BSE). (2016). *Bet (Bucharest exchange trading)*. Retrieved September 27, 2016, from http://www.bvb.ro/FinancialInstruments/Indices/Overview

Crișan-Mitra, C., & Borza, A. (2015). Approaching CSR in Romania: An empirical analysis. *Procedia - Social and Behavioral Sciences, 207*, 546–552.

Ecologia. (2011). *Handbook for implementers of ISO 26000, global guidance standard on social responsibility*. Middlebury Vermont: Ecologia. Retrieved September 18, 2015, from http://www.ecologia.org/isosr/ISO26000Handbook.pdf

Ernst & Young. (2015). *Trends and realities CSR in Romania*. Retrieved September 29, 2016, from http://www.ey.com/Publication/vwLUAssets/CSR_Survey_15_May_15/$FILE/CSR%20Survey%202015_RO.pdf

Ernst & Young. (2016). *Trends and realities CSR in Romania*. Retrieved September 29, 2016, from http://www.eyromania.ro/sites/default/files/attachments/EY_Studiu%20Tendin%C8%9Be%20%C8%99i%20realit%C4%83%C8%9Bi%20CSR%20%C3%AEn%20Rom%C3%A2nia%E2%80%932016_RO.pdf

Euractiv.ro. (2014). *Directiva votată în Parlamentul European/Impact în Romania: Raportare nefinanciară obligatorie pentru 720 de companii mari*. http://arhiva.euractiv.ro/index.html/articles%7cdisplayArticle?articleID=25970

European Commission. (2007). *Corporate social responsibility, national public policies in the European Union*. Brussels: Publications Office.

European Commission. (2014). *Directive 2014/95/EU of the European Parliament and of the council of 22 October 2014 amending Directive 2013/34/EU as regards disclosure of non-financial and diversity information by certain large undertakings and groups*. Retrieved September 15, 2016, from http://eur-lex.europa.eu/legal-content/EN/TXT/HTML/?uri=CELEX:32014L0095&from=EN

Filip, R., Iamandi, I.-E., & Munteanu, S. M. (2012). Corporate social responsibility management standards in Romania: Application and evaluation at corporate level. *Business Excellence and Management, 2*(2), 35–46.

Gilbert, D. U., Rasche, A., & Waddock, S. (2011). Accountability in a global economy: The emergence of international accountability standards. *Business Ethics Quarterly, 21*(1), 23–44. https://doi.org/10.5840/beq20112112

Global Reporting Initiatives (GRI). (2013). *GRI G4 guidelines part 1 reporting principles and standard disclosures*. Amsterdam: Global Reporting Initiatives. Retrieved September 15, 2016, from https://www.globalreporting.org/resourcelibrary/GRIG4-Part1-Reporting-Principles-and-Standard-Disclosures.pdf

Heal, G.M. (2004). *Corporate social responsibility - An economic and financial framework*. Retrieved February 26, 2015, from http://ssrn.com/abstract=642762. DOI: https://doi.org/10.2139/ssrn.642762

Iamandi, I. E., & Munteanu, S. M. (2011). Reporting about corporate social responsibility in Romania: Challenges and opportunities in an European context. In E. Druică (Ed.), *New challenges in economics and administration, conference proceedings* (pp. 360–374). Bucharest: Cartea Studențească Publishing House.

International Organization for Standardization (ISO). (2010). *ISO 26000 - guidance on social responsibility*. Geneva: International Organization for Standardization. Retrieved September 23, 2016, from https://www.iso.org/obp/ui/#iso:std:iso:26000:ed-1:v1:en

International Organization for Standardization (ISO). (2015). *Introduction to ISO 14001:2015*. Geneva: International Organization for Standardization. Retrieved September 27, 2016, from http://www.iso.org/iso/introduction_to_iso_14001.pdf

Kaphengst, T., Smith, L., Hohnen, P., Huetz-Adams, F., & Schneeweiss, A. (2013). *New options for strengthening standards on social and environmental responsibilities of corporations and their implementation*. Brussels: European Parliament, Directorate-General for External Policies, Policy Department. Retrieved September 25, 2016, from http://www.europarl.europa.eu/RegData/etudes/etudes/join/2013/457138/EXPO-DEVE_ET%282013%29457138_EN.pdf

Obrad, C., Petcu, D., Gherheș, V., & Suciu, S. (2011). Corporate social responsibility in Romanian companies - between perceptions and reality. *Amfiteatru Economic, 13*(29), 43–55.

Social Accountability International (SAI). (2014). *Social accountability 8000 standard*. New York: Social Accountability International. Retrieved September 22, 2016, from http://sa-intl.org/_data/n_0001/resources/live/SA8000%20Standard%202014.pdf

Szatmari, I., Macovei, C., Horia, C., & Aston, J. (2012). *Implementarea responsabilității sociale. Ghid de instrumente și tehnici*. Retrieved October 16, 2016, from http://undp.ro/libraries/projects/CSR/deliverables/Implementarea%20Responsabilitatii%20Sociale%20-%20Ghid%20de%20instrumente%20si%20tehnici.pdf

The Azores. (2016). *Romania CSR Index 2015*. Retrieved September 27, 2016, from https://www.slideshare.net/theazores/romania-csr-index-2015-56735169

Tschopp, D., & Nastanski, M. (2014). The harmonization and convergence of corporate social responsibility reporting standards. *Journal of Business Ethics, 125*, 147. https://doi.org/10.1007/s10551-013-1906-9

Waddock, S. (2008). Building a new institutional infrastructure for corporate responsibility. *Academy of Management Perspectives, 22*(3), 87–108. https://doi.org/10.5465/AMP.2008.34587997

Claudiu George Bocean, is a Professor and PhD supervisor at the Faculty of Economics and Business Administration within the University of Craiova. In 2007, he obtained his PhD degree in Economics within Faculty of Economics, University of Craiova, Romania and in 2015 he was awarded his Habilitation title in Management within Academy of Economic Sciences Bucharest, Romania. Since 2002 he has been teaching in the Faculty of Economics and Business Administration, University of Craiova taking courses in topics such as human resource management, labor economics, organization theory, business ethics, and has been co-operating in research projects with national and international universities and organizations.

Catalina Sitnikov, PhD is a Professor at University of Craiova in Romania, at the Faculty of Economics and Business Administration. She holds a Doctorate degree in Management since 2000, Habilitation title in Management since 2014 and since February 2015 been a PhD supervisor in Management. She was for 3 years a Visiting Lecturer at Helsinki University of Technology, Lahti Center in Finland. Since 1995, she has been teaching undergraduate, Master's and PhD students. She teaches Quality Management, Total Quality Management and Management. Her main research areas include: management, strategic management, and mostly quality management, instruments and models specific to the stages of quality planning, control and improvement, quality management strategies, ISO standards, CSR from the perspective of specific standards and instruments.

ISO 26000: A Brief Literature Review

Diana Iulia Popa and Dan-Cristian Dabija

1 Introduction

In the knowledge-based society, marked by social networks and rapid access to information, organizations are no longer concerned solely with making profit. As part of their general business strategy, they also include social as well as environmental protection objectives. A company can no longer advance without implementing social responsibility measures and activities, because these augment its performance and success with customers, who are more and more sensitive to societal and environmental issues (Debnath and Kamruzzaman 2015).

The increasing importance of environmental management for companies, along with their having to meet the environmental requirements and demands of the society in which they operate, as well as the demarcation of corporate social responsibility by highlighting its characteristics, examples of good practice and concrete actions to be taken has prompted the International Organization for Standardization to issue a number of specific standards over the last 20 years. Beginning with the ISO 14001 standard developed in 1996, more than 30 standards have been issued over time to regulate organizational operations according to environmental requirements, and to establish norms concerning product quality, labelling, etc. (Feldman 2012). Of notable importance are SA 8000, on the improvement of employee working conditions and AA 1000, which lays the foundation of the concept of sustainable development, etc. (Otova and Stankova 2011).

The ISO 26000 standard is intended for organizations that develop and implement economic, social and environmental sustainability activities. It provides a guideline which helps organizations to carry out their activities in compliance with

D. I. Popa · D.-C. Dabija (✉)
Babeş-Bolyai University, Cluj-Napoca, Romania
e-mail: iulia.popa@econ.ubbcluj.ro; cristian.dabija@econ.ubbcluj.ro

S. O. Idowu et al. (eds.), *ISO 26000 - A Standardized View on Corporate Social Responsibility*, CSR, Sustainability, Ethics & Governance,
https://doi.org/10.1007/978-3-319-92651-3_6

social and environmental requirements and also in harmony with their own individual priorities. Developed in 2005 and completed in 2010, the ISO 26000 standard was issued by the International Organization for Standardization (ISO) in order to enlarge the scope and applicability of the concept of Corporate Social Responsibility (CSR). Among the opportunities created by the implementation of ISO 26000 are: identification of ways to innovate a company, reduction in the risk of environmental pollution resulting from the company's activities, products or services and increased competitiveness of organizations enforcing the standard. The standard highlights seven key subjects underpinning the development and operation of any organization: corporate governance, human rights, working conditions, the environment, business practice, problems and challenges faced by consumers and, finally, social involvement (Ecologia 2011). Complex and yet simple, the ISO 26000 standard can be of great help to organizations everywhere willing to successfully implement the CSR concept and include it in their general business strategy.

Following presentation of the development of new standards by the International Organization for Standardization, the ISO 26000 standard is approached from a theoretical as well an empirical and practical standpoint. The third section of the paper discusses the relationship between Corporate Social Responsibility and ISO 26000, based on theoretical studies in the literature and empirical studies. The paper ends with the presentation of some relevant conclusions concerning the effects and prospects of developing the standard for organizations.

2 Developing ISO Standards

The new international standards were developed by a working group consisting of all stakeholders deemed important for the standard concerned. Created by ISO in 2005, this working group included more than 300 experts from over 54 countries where ISO standards were applied. The interested stakeholders and working groups participated in contextualizing and defining international standards according to the extent to which the implementation of such standards impacted upon their activity. International standards such as ISO 26000 are mostly applied within industrial organizations carrying out production activities. Government institutions are also interested in the adoption of international standards. The content of international standards, notably the way in which they are applied and contribute to the quality of life has attracted the interest of consumers as well. Consumers often direct their preferences towards companies (stores, producers, trademarks, products, etc.) which strictly abide by specific standards, thereby inspiring trust and guaranteeing the quality of products and services (Balzarova and Castka 2012).

In their study on the evolution of international standards, Balzarova and Castka (2012) analysed the influence of stakeholders on defining and developing standards according to five actions: eliminating and/or avoiding the unintended consequences of a phenomenon through standardization, connecting the new standards to previous versions and to other international norms and regulations specified in good practice guidelines, analyses, reports etc., encouraging consensus and dialogue between all the relevant stakeholders when developing a new standard, ongoing improvement of the content of standards for better application, and ensuring that the content of standards is periodically updated, expanded and improved. The working groups and relevant stakeholders to whom the standards apply pay increased attention to specific aspects, such as clarity of text, reduction of content, avoidance of any lack of clarity or non-compliance, elimination of sections which are useless or contradict recent regulations, provision of more examples of good practice, etc.

3 ISO 26000 Standard: Theoretical Versus Practical Dichotomy

The content of the ISO 26000 standard has been fully analysed and researched by many authors in the literature. Some studies take a holistic approach to the standard, while others deal with its constituent parts. Therefore, the first part of this study on the ISO 26000 standard contains the main premises of social responsibility, the second part explains the seven key areas of social responsibility and their characteristics, and the third part deals with the operationalization of social responsibility programmes, providing examples of good practice that can be applied in any sort of organization (Feldman 2012). The content and characteristics of the ISO 26000 standard are analysed from a theoretical and practical perspective in specialised reference works. A large number of studies highlight the major aspects of the ISO 26000 standard, pointing to its contribution to the development of specialised theory and practice as well as its advantages, disadvantages and influence on the sustainable development of companies applying it. The research also focuses on tailoring the standard according to fields of activity, industries, countries or groups of organizations. Other studies connect the standard to various theoretical concepts (value chain, social responsibility, etc.).

Hahn (2013) conducted a theoretical review on how the development of the ISO 26000 standard helps companies to implement successful social responsibility strategies, while Missimer et al. (2014) analyse the contribution of the standard to business sustainability by highlighting its benefits and shortcomings. The advantages, challenges and mechanisms for implementing the standard are analysed by Belaşcu et al. (2012), while Koszewska (2010) pinpoints the influence exerted by ISO standards on consumer choice when buying textiles and clothing items. In their literature review, Liew and Luetge (2016) describe the adoption of integrated management strategies by applying social responsibility within an organization's business practice, while Abrahamsson et al. (2010) highlight the advantages and disadvantages of integrated management systems. Abdirahman et al. (2014) develop a conceptual model for applying corporate social responsibility within small and medium-sized food stores, while Nunes (2017) focuses on the possibility of including elements of the ISO 26000 standard in social responsibility strategy. Bazillier and Vauday (2014) conduct a theoretical analysis of the concept of social responsibility based on the ISO 26000 standard and present its prospects for development.

Dankova et al. (2014) explain social responsibility standards from the perspective of the stakeholders of an organization, while Mazjin and Reveret (2015) deal with social responsibility pertaining to value chain in the field of social economy. Social responsibility based on the ISO 26000 standard is the subject of a paper by Castka and Balzarova (2008), as well as that of Mureşan (2016), who reveals the specific tools used by Romanian companies to target the market.

The ISO standard has been the object of many empirical investigations from an empirical and practical perspective. The research of Balzarova and Castka (2012) attempts to determine the extent to which organizations willing to implement ISO 26000 within their business strategy are actually successful in this respect. Hernaez et al. (2012) analyse the social responsibility activities of organizations from seven EU countries, establishing their real needs and problems, as well as the challenges they face when adopting customer-oriented specific measures. On the other hand, Formankova et al. (2017) refer to the most commonly used social responsibility standards in terms of their applicability in tertiary education. Cherapanukorn and Focken (2014) highlight the way in which Asian luxury hotels communicate their CSR policies, practices and strategies to target groups. Barnes and Croker (2013) analyse the principles and components of ISO 26000 that are most relevant to construction companies in Hong Kong, and offer them useful recommendations and good practice guidelines to improve their general business strategies and target group approach strategies. The qualitative research of Babin (2008) identifies the role of social responsibility in the decision-making process of some IT organizations. Diaz et al. (2013) pinpoint the internal and external ramifications of the concept of social responsibility in research centres and the nanotechnology industry. Feldman (2012) highlights how a company's actions, products and services can be re-tailored in line with the principles of environmental protection. Gueorguiev and Krasteva (2014) provide examples of the effects of implementing social responsibility measures by a Bulgarian energy company, and highlight positive aspects to help expand the customer segment, increase visibility and improve the company's image. Following their analysis of CSR methods applied by the Grameenphone company according to ISO 26000 principles, Debnath and Kamruzzaman (2015) deal with the impact of CSR strategies on the company's various stakeholders, and analyse in detail how the business relations with each of these stakeholders can be improved. Otova and Stankova (2011) investigate the extent to which European companies in all sectors are prepared to implement the ISO 26000 standard within their business strategies, and reach the conclusion that there is still potential for better and more careful use of such strategies. Likewise, Persic and Markic (2013) conducted a quantitative study measuring the impact of social responsibility on company operations, with special focus on the ways to foster success. As there is no specific scale for the empirical measurement of ISO 26000, Chow et al. (2016) developed a possible method for measuring the standard which they recommend for use within quantitative research.

Castka and Balzarova (2008) believe that this standard was issued to support company development, increase customer satisfaction with and trust in products and services, and to complement previous ISO standards, such as ISO 14001, ISO 900 and ISO 14000. Balzarova and Castka (2012) note that, unlike previous standards, ISO 26000 has a broader scope of applicability, with special focus on social responsibility, which can be applied to any type of organization. Bazillier and Vauday (2014) claim that the ISO 26000 standard is a useful good practice guideline proposing an international definition of social responsibility that can be used by any company. Going back to the last century, environmental management has been of great interest to entrepreneurs and business practices everywhere, and is governed by the ISO 14001 standard issued in 1996. The clear regulation and explanation of environmental protection activities by organizations of any size (Feldman 2012) has been accomplished over time by means of several specific standards, such as ISO 900 (quality management) and ISO 14000 (environment). Unlike these two, ISO 26000 is a generally accepted standard, applicable to any field by any type of organization (Castka and Balzarova 2008). It is actually the only internationally valid norm which does not entail the existence of a certification, and is based on an essential characteristic: its voluntary applicability by organizations (Corporate Excellence 2011). Otova and Stankova (2011) state that the fundamental purpose of the ISO 26000 standard is to regulate the concept of corporate social responsibility at an international level and make it more attractive to organizations. By applying the principles of the standard, organizations may increase their recognition and visibility among customers, which leads to greater competitive advantage and better differentiation from competitors.

Hahn and Weidtmann (2012) claim that the development of the ISO 26000 standard has passed through four stages: the preliminary stage, covering a feasibility study and the need to define it; consensus within the working committee on its contents and structure; researching its effects; followed by the stage in which the final form of the document was approved. These four stages provide clear evidence that the ISO 26000 project was based on the desire to reach a broad consensus between all those involved, and between those affected by the standard.

Being intended for organizations of any kind, ISO 26000 can easily be implemented by both private and public companies, and developed and developing companies adopting social responsibility for the first time, as well as by experienced companies that only use this strategy (Mureşan 2016). Some sources (Hahn 2013) recommend the application of ISO 26000 by companies using the standard for the first time.

4 CSR and ISO 26000: Theoretical and Practical Approaches

4.1 Conceptual Demarcations of Social Responsibility in Light of ISO 26000

The concept of CSR is increasingly popular among organizations from various sectors and countries, and has been dealt with at great length by theoreticians and practitioners. Depending on how it is actually implemented, there have been different definitions and characterizations of the concept in the literature. One definition of the concept of social responsibility pertains to the responsibility assumed by organizations in making corporate decisions, especially when such decisions take into account the impact of their activities on the environment and society in which the company operates with production, supply, marketing units, etc. Naturally, an organization has to adopt ethical behaviour whenever it implements social responsibility actions (Corporate Excellence 2011). A similar approach is offered by Abdirahman et al. (2014). According to them, CSR is a concept adopted by modern companies to keep up with new trends in society that compel them to take into account social requirements and opportunities to protect the environment. For Otova and Stankova (2011), CSR is about implementing transparent business strategies based on ethical values and developed in compliance with the legal requirements and regulations of each country and/or region. It takes into account each community in which the company operates, and the rigour associated with environmental protection and preservation. The interest of practitioners and theoreticians in social responsibility has increased substantially over the past few years. This trend is expected to continue in the future (Babin 2008). The great interest in environmental issues and ways of protecting the environment, together with concrete resource preservation actions and the development of internationally recognized and accepted social responsibility standards such as ISO 26000, are proof that the new principles are being adopted by an increasing number of public and private organizations. It is the author's opinion that the popularity and applicability of CSR actions is more and more important and relevant to all stakeholders of a company. Where companies are interested in making a quick profit, consumers are likely to become more and more sceptical of the way in which they carry out actual and efficient social responsibility actions and adopt correct measures to protect the environment. This will encourage consumers to become more involved in CSR actions (Bazillier and Vauday 2014), punishing companies (for example, by boycotting brands or products) in the case of failure to live up to the promised standards (Dabija et al. 2016). The development of international standards has made a great contribution to improving the exchange of goods between regions and companies, aiding overseas investments, spreading information and examples of good practice in social responsibility and environmental protection, and sharing practices on the use of various technologies within production and management processes (Balzarova and Castka 2012). CSR actions are becoming more important not only to companies, but also to consumers who increasingly assess organizations on the extent to which they adopt such a strategy and exhibit ethical behaviour when developing, distributing and marketing various products and services (Koszewska 2010). Social actions are, therefore, a means of dealing with various prevalent social and environmental issues, such as pollution, global warming, poverty, illiteracy, etc. (Dabija et al. 2016).

5 CSR and ISO 26000 from Empirical and Applicative Sectorial Perspective

In addition to the study of company efficiency and involvement in CSR actions, the literature also identifies company success in implementing the ISO 26000 standard. Based on their quantitative study on 135 Slovenian organizations from 11 different sectors, Persic and Markic (2013) conclude that the implementation of CSR actions was part of each company's vision and strategy, designed so as to cover the impact of their decisions and actions on the environment and society. The adoption of such vision and strategy led to better and more effective communication between employees and consumers, a greater volume of investments and better qualified/trained employees. In most cases, the application of the ISO 26000 standard was directly proportional to the size, income and profit of each organization. Appropriate planning and the implementation of strategies in compliance with CSR principles led to more positive results for each company, ensuring long-term sustainable development. Despite the Slovenian entrepreneurs' scant knowledge of the content and application of the ISO 26000 standard, they showed great interest in social responsibility and principles, which further increased the impetus for adopting the ISO 26000 standard within the studied organizations (Otova and Stankova 2011). We believe that the ISO 26000 standard and CSR actions can be included within the practices and/or general business strategy for every company, regardless of its sector.

Chow et al. (2016) attempted to develop a quantitative scale for measuring the degree to which ISO 26000 was applied within 286 companies in Hong Kong, based on the seven key principles of the standard. Their conclusion was that the most popular CSR practices adopted by organizations related to environmental protection, the improvement of employees' working conditions and the mitigation of social problems. Protecting employees' rights is an essential requirement for modern companies. In so doing, companies prove their involvement in internal CSR activities (towards employees) and are able to attract and maintain the loyalty of a highly qualified labour force. The external social responsibility of the companies investigated concerned the social problems of the communities in which they operated, supporting financially various charities and performing concrete actions for this purpose. Company involvement in solving environmental issues represented a strategy that strengthened their competitive position and yielded significant benefits by reducing overhead expenses (consumption of energy, water, etc.). Finally, the study showed that companies were also concerned, albeit at a lower level/intensity, with protecting the rights of consumers by providing healthy, safe products, brands and services produced according to ethical norms (Chow et al. 2016).

Some content of the ISO 26000 standard was irrelevant to the field of construction services in Hong Kong (Barnes and Croker 2013). However, the aspects concerning social responsibility, employee working conditions and customer problems were best dealt with by companies with over 200 employees. The greater the number of employees of the investigated companies, the better they complied with the principles of the ISO 26000 standard. Customers' rights, respecting employees' working conditions and environmental management were issues closely scrutinized by companies having between 50 and 100 employees.

A large number of Asian luxury hotels have adopted ISO standards, notably ISO 26000, as reference for their own CSR activities (Cherapanukorn and Focken 2014). There were, however, various types of tourist accommodation where the principles of ISO standards were not very clear. Many long-entrenched business practices were erroneously viewed as CSR actions, even if most managers of the investigated Asian hotels said that CSR actions were essential for the development and strengthening of their own business. Similar results were obtained by Krukowska (2014), who conducted secondary research on the factors determining Japanese companies to carry out social responsibility activities. She states that there were four categories of factors driving CSR actions: cultural and religious factors, traditions and values; legislative regulations; public opinion; historical events and traditions. These results are strongly correlated with Japanese business philosophy. For the Japanese, culture, religion, social values and traditions are essential norms of daily life and influence the involvement of companies in CSR activities. However, these factors contrast with Western values which adopt a universal and broad approach to the principles of the ISO 26000 standard. This highlights the difference between Japanese companies and European or American companies in approaching social responsibility (Fukukawa and Teramoto 2009; Krukowska 2014). The importance of social responsibility increased in the case of telecommunication services, as companies carried heavier responsibility to ensure the security of information sent to customers. Therefore, observing the code of business ethics along with CSR norms and the development and strengthening of new strategies in the era of social media and digital communication is essential in maintaining the balance (Debnath and Kamruzzaman 2015).

In countries such as Slovenia, Poland, Lithuania and Bulgaria, social responsibility is ignored in the educational system. The curriculum only includes certain CSR aspects, such as environmental protection, but excludes the other key topics included in ISO 26000 (Hernaez et al. 2012). After conducting an analysis of six international standards and two national standards applied in higher education institutions in the Czech Republic, Formankova et al. (2017) concluded that ISO 26000 was the clearest, most logical and comprehensive standard, and recommended it as the basis for developing the social responsibility policies of these institutions. Similar conclusions were reached by Gueorguiev and Krasteva (2014). Based on a case study conducted in a Bulgarian hydro-electric power station, Vaptech and in the University of Ruse, the authors concluded that social responsibility was broadly applied both locally and globally, and enhanced the popularity of the companies in supporting the development of their local communities.

In their study of the CSR concept in nanotechnology centres, Diaz et al. (2013) noted a different approach to this concept, namely, with reference to the protection of staff working in these centres. Such research centres hired scientists who exchanged information with other similar national and international institutions. In the authors' opinion, it was essential to study the risks faced by these centres, as there was a need to anticipate the potential negative effects caused by the obtained results. They also stressed the usefulness of developing effective communication between all stakeholders in these research centres, and having dialogue with local communities to better understand the impact of the results on their development, because the most important mission of social responsibility was gaining society's trust in the activities of the research centres. Approaching the CSR concept was very important for improving both customers' and investors' trust in the studies conducted by specialized nanotechnology centres.

Social responsibility and ISO 26000 do not always complement each other. There are cases in which the statements of companies concerning specific measures to be implemented do not match the real facts. This may quickly diminish or even lose consumer trust in the brands, products and services of such organizations. A telling example comes from the textile and clothing industry (Koszewska 2010) where there is danger of exploiting the labour force, producing clothes at very low cost in least developed countries, polluting the environment during production or transport processes, etc. (Dabija et al. 2016). The value chain in this sector is relatively long and vulnerable, making the effective control of each stage and of all stakeholders involved even more difficult. Therefore, urging all participants to adopt and comply with all stages of the ISO 26000 standard would guarantee the best results concerning social responsibility and business ethics, increasing the trust and satisfaction of the end customer.

6 Conclusions

The concept of CSR is undergoing an ongoing evolution at an international level, as companies in most economic sectors have adopted and implemented it. Among the main reasons behind CSR actions is the need to improve an organization's financial success and, in particular, to capture the customer's attention, invoking satisfaction and loyalty. Company involvement in CSR actions has stirred interest, and become more and more important in the consumer's decision-making process when purchasing a product, service or brand.

The literature review shows that companies, as well as various public institutions are increasingly adopting social responsibility actions and strategies. Ensuring a high degree of uniformity in social responsibility measures, tactics, good practice examples, etc., is now possible, thanks to the ISO 26000 standard. This standard contains internationally valid definitions, principles, characteristics and values, and, by popularizing social responsibility, companies everywhere benefit considerably by implementing CSR actions. By adopting the standard, even partially, organizations can increase their international visibility and trust, gaining valuable partners and new customers.

The degree to which the ISO 26000 standard is applied varies according to the size of organization, being mostly adopted by large and high profit companies. The literature indicates many examples of good practice in the private as well as the public sector in hotel, tourist, educational, construction and production services. Naturally, its seamless adoption still poses many difficulties and challenges, but studies have shown that company management has made great efforts to obtain the best effects. The inclusion of international standards on sustainability, environmental management, quality management, social responsibility, etc., within general business strategy enables companies to gain competitive advantage, improving their reputation and international visibility, strengthening relationships with stakeholders, boosting employee morale and trust, and retaining qualified manpower (Debnath and Kamruzzaman 2015). Consequently, the success derived from implementing such standards will not be long in coming.

Acknowledgment This work was supported by a grant of the Romanian National Authority for Scientific Research and Innovation, CNCS-UEFISCDI, project number PN-II-RU-TE-2014-4-0312.

References

Abdirahman, Z. Z., Suavee, L., & Shiri, G. (2014). Analysing network effects of corporate social responsibility implementation in food small and medium enterprises. *Journal on Chain and Network Science, 14*(2), 103–115.

Abrahamsson, S., Isaksson, R., & Tosteby, J., (2010). *Integrated management systems: Advantages, problems and possibilities*. In: J. Martin (Ed.) 13th Toulon-Verona conference "Organisational excellence in services". Coimbra, pp. 567–578.

Babin, R. (2008). Assesing the role of CSR in outsourcing decisions. *Journal of Information Sysmes Applied Research, 1*(2), 3–14.

Balzarova, M. A., & Castka, P. (2012). Stakeholders' influence and contribution to social standards development: The case of multiple stakeholder approach to ISO 26000 development. *Journal of Business Ethics, 111*(2), 265–279.

Barnes, L. R., & Croker, N. (2013). The relevance of the ISO26000 social responsibility issues to the Hong Kong construction industry. *Australasian Journal of Construction Economics and Buiding, 13*(3), 37–50.

Bazillier, R., & Vauday, J. (2014). CSR into (new) perspective. *Foresight, 16*(2), 176–188.

Belaşcu, L., Baltador, L., Budac, C. (2012). *ISO 26000 – Benefits, difficulties, questions*. Proceedings of the international scientific conference, Eco-Trend 2010 (pp. 675–681). Târgu-Jiu: House Publisher Academia Brâncuşi.

Castka, P., & Balzarova, M. A. (2008). ISO 26000 and supply chains – On the diffusion of the social responsibility standard. *International Journal of Production Economics, 111*, 274–286.

Cherapanukorn, V., & Focken, K. (2014). Corporate social responsibility (CSR) and sustainability in Asian luxury hotels: Policies, practices and standards. *Asian Social Science, 10*(8), 198–209.

Chow, P.-S., Tang, A. K. Y., & Yip, A. C. Y. (2016). Scale development and operationalization of social responsibility constructs: An ISO 26000 context. *Business and Economic Research, 6*(2), 156–175.

Corporate Excellence. (2011). *A new common framework for corporate social responsibility: The international standard ISO 26000*. Retrieved October 29, 2017, from http://www.corporateexcellence.org/index.php/eng/Sharing-Knowledge/A-New-Standard-for-CSR-ISO-26000
Dabija, D. C., Pop, N. A., & Postelnicu, C. (2016). Ethics of the garment retail within the context of globalization and sustainable development. *Industria Textilă, 67*(4), 270–279.
Dankova, P., Valeva, M., & Strukelj, T. (2014). A comparative analysis of international corporate social responsibility standards as enterprise policy/governance innovation guidelines. *Systems Research and Behavioral Science, 32*(2), 152–159.
Debnath, G. C., & Kamruzzaman, M. (2015). Implementation of 7 core principles of ISO 26000 (CSR) in Bangladesh: A case study on Grameenphone. *Daffodil International University Journal of Business and Economics, 9*(1), 173–185.
Diaz, E. C., Valcarcel, M., & Simonet, B. M. (2013). The social responsibility of nanoscience and nanotechnology: An integral approach. *Journal of Nanoparticle Research, 15*, 1–13.
Ecologia. (2011). *Handbook for implementers of ISO 26000, global guidance standard on social responsibility*. Middlebury Vermont: Creative Commons.
Feldman, I. R. (2012). ISO standards, environmental management systems, and ecosystem services. *Environmental Quality Management, 10*, 69–79.
Formankova, S., Hrdlickova, A., Kucerova, R., & Skypalova, R. (2017). Evaluation of the applicability of CSR standards in the praxis of tertiary education institutions. *Forum Scientiae Oeconomia, 5*(1), 95–103.
Fukukawa, K., & Teramoto, Y. (2009). Understanding Japanese CSR: The reflections of managers in the field of global operations. *Journal of Business Ethics, 85*, 133–146.
Gueorguiev, T., & Krasteva, M. (2014). Social responsibility as a means for sustainable development in the lower Danube River basin. *Journal of Danubian Studies and Research, 4*(1), 52–62.
Hahn, R. (2013). ISO 26000 and the standardization of strategic management processes for sustainability and corporate social responsibility. *Business Strategy and the Environment, 22* (7), 442–455.
Hahn, R., & Weidtmann, C. (2012). Transnational governance, deliberative democracy, and the legitimacy of ISO 26000: Analysing the case of a global multistakeholder process. *Business & Society, 55*(1), 90–129.
Hernaez, O., Zugasti, I., Waltersdorfer, G., Matev, D., Assenova, M., Jonkute, G., Staniskis, J., Schoenfelder, T., Bogataj, M., Moller, J.D., Hirsbak, S., Schmidt, K., Christiansen, K., Fondevila, M., & Arenda, J. (2012). *Corporate social responsibility on SMEs*. 15th European roundtable on sustainable consumption and production. Bregenz, Austria.
Koszewska, M. (2010). CSR standards as a significant factor differentiating textile and clothing goods. *Fibres and Textiles in Eastern Europe, 18*(6), 14–19.
Krukowska, M. (2014). Determinants of corporate social responsibility in Japanese companies. *Management and Business Administration. Central Europe, 22*(4), 39–57.
Liew, P. Y., & Luetge, C. (2016). Integrated management systems frameworks for corporate social responsibility and related concepts. *Journal of Management and Sustainability, 6*(3), 12–24.
Mazjin, B. & Reveret, J.-P., (2015). Life cycle sustainability assessment: A tool for exercising due diligence in life cycle management. In: G. Sonnenmann, M. Margi (Eds.), *7th International conference on life cycle management (LCM)* (pp. 51–63). Bordeaux: Springer.
Missimer, M., Robert, K.H., Broman, G. (2014). A systems perspective on ISO 26000. In *2nd international symposium "Systems thinking for a sustainbable economy. Advancements in economic and managerial theory and practice"*. Rome: Univesitas Mercatorum.
Mureşan, L. (2016). Ethical instruments for the development of the social responsibility of the business operators in Romania. *Bulletin of the Transylvania University of Braşov, 9*(58), 65–72.
Nunes, C. C. (2017). Social responsibility and ethical decisions - a presentation of elements contained in ISO 26000:2010, guidance on social responsibility. *Quality – Access to Success, 18*(158), 42–44.

Otova, M. & Stankova, E., (2011). The preparedness of certain companies to implement the ISO 26000 standard. In: Kocourek, A., Ortova, M., Mysakova, D. (Ed.), *Proceedings of the 10th international conference. Liberec Economic Forum 2011* (pp. 372–381). Liberec: Technical University of Liberec.

Persic, A., & Markic, M. (2013). The impact of social responsibility vision and strategy on successful corporate operations. *Managing Global Transitions, 11*(1), 27–40.

Diana Iulia Popa, is a PhD candidate at the Department of Marketing, Faculty of Economics and Business Administration, Babeş-Bolyai University, Cluj-Napoca, Romania. She obtained her bachelor's degree in 2015 and her Master's degree in 2017, both at the Faculty of Economics and Business Administration, Babeş-Bolyai University, Cluj-Napoca, Romania. In 2017 Diana started the PhD program. Her dissertation research focuses on omni-channel retail and experience marketing. She is also interested in sustainability as well as organic production and consumption.

Dan-Cristian Dabija, PhD, is an Associate Professor at the Department of Marketing, Faculty of Economics and Business Administration, Babeş-Bolyai University, Cluj-Napoca, Romania. His teaching and research fields are in retailing, international marketing, consumer behaviour and tourism marketing. Dr. Dabija completed his PhD Studies at the Bucharest University of Economic Studies, Romania, and has been awarded several doctoral and postdoctoral research scholarships at Romanian, Polish, British, Finisch, Hungarian and German universities. Dr. Dabija has lead a research project on sustainability in retailing financed by the Romanian National Authority for Scientific Research and Innovation. He has published in various journals and books and has participated in numerous conferences (EMAC, World Marketing Congress, Annual North American ACR Conference, RESER). He also serves as a reviewer for *Amfiteatru Economic*, *Marketing from Information to Decision, Marketing Education Review*, and *Management & Marketing Journal*.

Part II
Standardised CSR Cases on ISO 26000

ISO 26000 in Corporate Sustainability Practices: A Case Study of the Forest and Energy Companies in Bioeconomy

Anne Toppinen, Mirja Mikkilä, and Katja Lähtinen

1 Introduction

In the European Union (EU), the major challenge related to development of bioeconomy based on the usage of renewable resources is how to materialize a transformation towards a low-carbon, resource and energy efficient economy (Hetemäki et al. 2014). The utilization of forest resources for renewable energy production has been seen, for example, as an option to mitigate climate change, to enhance positive social development, and to support the business opportunities of forest industry companies of different sizes also in rural areas (Stupak et al. 2007). In line with this, various policy instruments have been introduced to enhance the renewable energy production in the EU since the early 2000s (Ericsson et al. 2004).

The global, traditional forest based bioeconomy (i.e., wood, pulp and paper production) is characterized with high capital intensiveness, mature markets of several core products, low innovation intensity and increasingly international firms operating in global markets with high price volatility (Pätäri et al. 2016). Through increased market globalization, a growing awareness of requirements for sustainability in business operations and a shifting of production capacity to low-income countries in the Global South, the forest based bioeconomy in Europe has also

A. Toppinen (✉)
Department of Forest Sciences, University of Helsinki, Helsinki, Finland
e-mail: anne.toppinen@helsinki.fi

M. Mikkilä
Lappeenranta University of Technology, Lappeenranta, Finland
e-mail: Mirja.Mikkila@lut.fi

K. Lähtinen
Department of Marketing, University of Vaasa, Vaasa, Finland
e-mail: klahtine@uwasa.fi

S. O. Idowu et al. (eds.), *ISO 26000 - A Standardized View on Corporate Social Responsibility*, CSR, Sustainability, Ethics & Governance,
https://doi.org/10.1007/978-3-319-92651-3_7

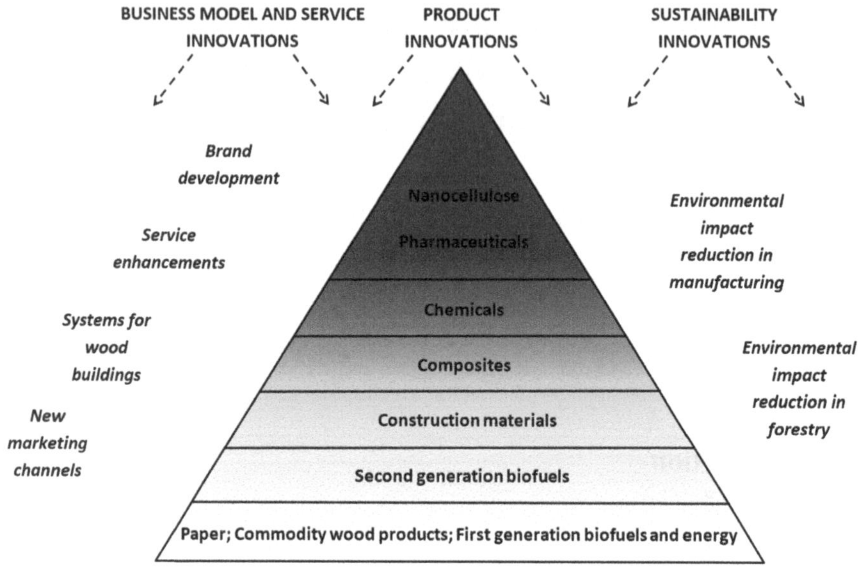

Fig. 1 Breakdown of different types of innovations in the foreseen development of bioeconomy value pyramid

become more exposed to growing vulnerability in competitiveness and company sustainability image (Mikkilä and Toppinen 2008). The overall changes in the global business environment of forest based bioeconomy call for realizing the necessary new green innovations amidst a prolonged global recession, especially in paper and wood products (Panwar et al. 2012).

In the ongoing process of creative destruction towards bioeconomy, sectorial boundaries are becoming blurred across in the intersections of different businesses in forest and energy industries, for example. For both industries, the sustainability impacts of renewable energy production vary depending on the origin of wooden raw material (e.g., domestic or imported) and the end-products processed (e.g., heat, electricity, solid or liquid biofuels) (e.g., Myllyviita et al. 2013). All transmittable and transportable renewable energy products (e.g., electricity, solid biofuels, liquid biofuels) may be sold both at local, regional and global markets, while heat is always being used locally as a result of cooling. Thus, it is not possible unambiguously define the social responsibility (SR) impacts of any business operations of forest and energy industries without consideration of their value chain structures.

Figure 1 illustrates how the boundaries of forest, energy and chemical industries are crossing in the bioeconomy to create products with higher value added, and what is the role of different types of innovations in the systemic change changing traditional division of industrial boundaries. The general vision (TEM 2014) seems to focus in aiming to move from the bottom of pyramid towards higher value added products and services. However, it is elementary to understand the systemic nature of changes required in industries and many interlinkages between scales of production between different types of products. While in some assessments it is possible to distinguish product and service innovations from business model innovations and from sustainability related (environmental) innovations, it is fair to say that the main body of ongoing innovation efforts towards bioeconomy have some reference to sustainability. Furthermore, most commonly sustainability innovation efforts are made from the resource or energy efficiency point of view or for substituting fossil-based materials with renewables.

Due to the fact of differences in the characteristics of different renewable energy products together with increasing integration of renewable energy production into the forest based bioeconomy, new ways of assessing, monitoring and standardizing social responsibility practices are evolving both at organizational and systemic level. In a recent review (Toppinen et al. 2016), advancing corporate sustainability thinking into a more systemic level was identified to require: (1) the extension of product-level thinking on value creation to recycling and disposal stages and (2) the incorporation of byproducts and investments into flexible production systems to produce both inter-firm and inter-sectorial synergies.

Within the systemic level developing framework forest bioeconomy and energy companies operate globally. Hence, it is to investigate whether the sustainability agenda of the selected companies meets the internationally, generally acceptable criteria, such as ISO 26000, and what is the value add of the application of the criteria for the companies.

2 Research Design

The role of ISO 26000 in corporate sustainability practice within forest bioeconomy and energy sectors were analyzed qualitatively within the framework of ISO 26000, following the analysis path by for example Miles and Huberman (1994) and Yin (2003). Here, especially the case study strategy by Yin's (2003) was perceived applicable in that we described organizational phenomenon within ISO 26000 framework. We chose four case companies; two companies, Metsä Group and Koskisen, representing forest bioeconomy and the other two cases, Fortum and Gasum, energy sector.

The applied data was secondary, publicly available material, such as newspaper articles, previous researches, CR and sustainability reports of the selected companies, and websites of international organizations in order to keep our analysis repeatable and the consequent findings transparent. The chosen material was reviewed within the ISO 26000 framework. In other words, we looked, firstly, direct references of the application of ISO 26000 by the analyzed companies. Secondly, we reviewed concepts and themes that referred ISO 26000 theme if no direct application was found.

Due to the limited number of cases and the related company originated material we handled the data manually, as on one hand the application of some data processing tool had not increased the efficiency and reliability of the data processing significantly and on the other hand the analysis path stays transparent and can be repeated due to the documentation of the manual process.

In later parts of this chapter we make an overview of current state and implementations of seven ISO 26000 core topics and their coherence with current sustainable business practices within forest based bioeconomy and energy industries from the perspective of Finland. Related to that, our first objective is to make a comparative assessment of sustainability implementation and analysis of ISO 26000 in corporate sustainability practices across selected four case companies. Second, we aim at scoping the challenges that organizations face in promoting a standardized view of their SR, especially from upstream sourcing of raw material. Third, we make conclusions and recommendations for developing practical sustainability management practices in the crossroads of forest based bioeconomy and energy industries.

3 ISO 26000 Standard in the Finnish Business Context

A voluntary ISO 26000 standard, introduced in 2010, provides guidance on the integration of social responsibility into management processes, as well as on the principles of SR. When adopting this standard, companies and other organizations can effectively translate their sustainability principles related to social responsibility into effective actions and best practices via employment of appropriate criteria and indicators (e.g., Toppinen et al. 2016). A recognition of a business enterprise' stakeholder is an important element in ISO 26000, as is the short- and long terms objectives in terms of pursuing SR. From the corporate responsibility perspective, growing public awareness in critical corporate conduct has placed an ever greater need to build and secure the legitimacy of operations through improved transparency, maintain trustful stakeholder relations and leverage the social capital inherent in these relations, so as to enhance value creation during increasing global competition.

ISO 26000 standard aims at providing information on understanding the role of sustainability in different types and sizes of organizations, helping to find practices for integrating the standards throughout the organization and communication on sustainability issues, thereby independently improving organizational practices related to SR. According to Hahn (2013), holistic ISO 26000 is useful in providing starting point for implementing organizational sustainability strategies and helpful in conducting internal and external analyses. The list of seven core subjects of ISO 26000 (i.e. organizational governance, environment, human rights, fair operating practices, labor issues, consumer issues and community involvement and development) presents the most essential areas of SR that an organization should consider to maximize its contribution to sustainable development (ISO 2012). In addition, each organization should actively recognize and address those areas in ISO 26000 that are most relevant to its own field, which calls for sector specific approach in assessing implementation of ISO and for identifying the related challenges. Although economic issues are not directly present, they are covered throughout all these seven dimensions.

The development process of ISO 26000 standard was very long and it has been also criticized (see e.g., Balzarova and Castka 2012; Marques 2012), based on that a lengthy multi-stakeholder process did not necessarily ensure legitimacy or guarantee that this non-certifiable standard could be considered as an enforceable instrument. Rasche (2010) also points out that responsibility standard alone can never be a complete solution to the plethora of social and environmental problems experienced today. According to Hemphill (2013), ISO 26000 seems for smaller-sized companies as a lengthy, complex document, which has a relatively high learning curve cost regarding its implementation of policies and practices. Furthermore, because it neither provides detailed guidance for implementing operational measures in the industry or sector context nor is certifiable, it fails in providing assurance for legitimacy. Castka and Corbett (2016) also argue that sustainability standards will be more widely adopted if they are better-governed, less stringent and more favorably covered in the media, and regarding ISO 26000 there seems room for development in all these areas.

The adoption of ISO 26000 within the Finnish business environment has been so far a smooth but a rather low-key process. Regardless of the intensive responsibility debate in the Finnish society in the 2000s, there is no corresponding policy on corporate responsibility at the national level unlike in the European Union (Parliament of Finland 2014). The main reason for this might be the welfare state status and advanced social legislation that have created an adequate framework guaranteeing minimum social services to the citizens and reasonable business environment for the private sectors. Also the voluntary nature of corporate responsibility as a set of actions that go beyond the legal obligations explains the low number of policy initiatives during the last decades (Mikkilä et al. 2015).

The Finnish business organizations can be categorized into three types: export-oriented large-scale industries, traditional or home-market operating small and medium-scale enterprises (SME) and newly established SMEs based on the commercialization of an innovative business idea. The financial, social and environmental operating environments of these actors vary significantly from each other, leading to various responsibility focuses. The Finnish large-scale companies integrate corporate responsibility dimensions in the operations through environmental and social management systems such as Environmental Management Auditing Scheme, EMAS (European Union EU 2009); Occupational Health and Safety Assessment Series, OHSAS (British Standard Institution, BSI 2007) and Social Accountability 8000, SA8000 (Social Accountability International, SAI 2008). The application of the systems is typically accredited through international standards (Mikkilä et al. 2015).

For the time being, Finland follows the practice of the majority of the other European Union countries in the voluntary corporate responsibility reporting. The Parliament has not initiated to enlarge the legal demand of annual reporting to cover also a larger set of social and environmental indicators. Global Finnish companies have started to standardize their responsibility reporting in the mid-2000s by adopting the Global Reporting Initiative (GRI) reporting format (GRI 2014). The context-dependence of responsibility documentation, especially reporting needs on SR of forest-based renewable energy production has been found to be vary notable in different situations among Finnish businesses (Myllyviita et al. 2013). For example, while reporting on non-usage of child labor and education opportunities may be crucial for assessing social sustainability, monitoring those issues is not relevant in areas like Finland without usage of child labor and with compulsory 9 year school education.

In the previous studies, sustainability managers in the North American and European companies operating within forest based bioeconomy companies have been found not yet to have a deep familiarity with the standard, and the practical benefit for the sustainability frontrunners of adopting the ISO 26000 have been questioned (Toppinen et al. 2015a). The managers also argued that a global and uniform social responsibility standard might not be sufficient for developing a more in-depth company-specific conceptualization of corporate responsibility, and that there is a risk that such a standard could become either too imposing or too superficial, and not able to address the sector specific issues. Furthermore, while the ISO 26000 guide on social responsibility has been found applicable as a schematic framework for analyzing sustainability communication content in the pulp and paper businesses, the emphasis on environmental sustainability seems to dominate in communication over issues (Toppinen et al. 2015b). In their case analysis of Finnish forest, mining and food sector companies, Mikkilä et al. (2016) found that a ISO 26000 social responsibility guideline provides a relatively comprehensive framework for the implementation of corporate responsibility, but at the same time the detailed revision of the cases indicated the technical-social focus of the ISO criteria set giving less consideration on environmental issues that are commonly perceived as one of the most relevant dimension of comprehensive responsibility in the Finnish context.

4 Results

4.1 ISO 26000 Within Forest and Energy Industries

Finland has been chosen as focus area due to national interests in promoting European climate strategy in its energy policy and national renewable energy program. The Finnish national Energy and Climate Strategy was updated in 2013 and its energy policies are well integrated with those of the European Union. The focal points of the government's energy strategy are to strengthen its energy security, to move progressively towards a decarbonised economy, and to deepen its integration in the wider European market (TEM 2016). Regardless of the country's high dependence on imported fossil fuels, decarbonising is the long-term objective, Finland having already one of the lowest shares of fossil fuels in its energy mix among IEA member countries, ranking fourth-lowest in 2011 (IEA 2013).

Finland has a very ambitious renewable energy program, with a view to meeting its binding EU target to increase the share of renewable energy to 38% of final energy consumption by 2020. The government has clearly indicated that forestry will play a central role in meeting its renewables target, with the sector having to contribute half of the additional 38 tWh between 2005 and 2020. Measures implemented to attain the country's renewables target include promoting the use of forest chips and other wood-based energy, alongside wind power, the use of biofuels in transport, and the greater utilization of heat pumps. Although the government is in favor of the requirement that biomass use be sustainable, there are serious concerns about potential EU schemes in this regard, which could bring about a great deal of administrative burden for their certification (IEA 2013).

One of the recent key approach in sustainable business models is the concept of circular economy introduced by the Ellen MacArthur Foundation in the early 2010s. The model is based on the philosophy of redefining products and services to design waste out, while minimizing negative impacts. Consequently, in a transition to renewable energy sources, the circular model builds economic, natural and social capital.

Finland is an example where the energy policy and the role of renewables in it was interlinking the traditional forest and energy sectors into circular economy type solutions before the debate itself was launched. The pulp and paper industry has a significant role in the Finnish renewable energy production due to the widely adopted kraft pulp production process since the 1930s. The by-product of the process, black liquor can be processed further as steam and electricity to be utilized in the pulp and paper process or to feed the electric grid. In addition, the pulp and paper processes cogenerate electricity and heat within using biomass-based by-products and wastes as fuel. As a consequent, the share of forest sector in national bioeconomy output in Finland is as high as 50%, and that of the energy supply 11% totaling in two-thirds of the GDP. Furthermore, the pulp and paper industry corresponds for 25% of the total electric production equaling 70% of the renewable energy production in Finland due to energy production within the pulp and paper process (Forest Industries 2017; TEM 2014). Furthermore, the forest based bioeconomy companies have had co-projects with the energy companies in biofuels production (see e.g. Stora Enso and Neste Oil company websites).

Finland has a high level of energy consumption per capita within European Union due to the cold climate, long distances and relatively energy-intensive industry structure (Findicator 2017). Regardless of the private consumption habits one of the major cause is the large pulp and paper industry basis with its relatively energy intensive production processes even though the processes themselves produces a significant share of the energy produced in Finland. The main target of the energy companies has traditionally been the energy supply for the households and industries, but recently the energy companies have started to profile themselves more and more as circular economy specialists in processing waste, heat and energy—one example of this being a large-scale energy player Fortum Oyj that acquired a circular economy company Ekokem in 2016 (Fortum Oyj website).

5 Overview of Case Companies and Their Stance Towards ISO 26000

In the following we will first make an overview of the four case companies and the role of ISO 26000 core topics present in their management systems (Tables 1 and 2). The data are extracted from their most recent corporate sustainability reporting, web-pages and other publicly available documents. The analysis aims at elaborating similarities and differences across businesses and, consequently, pointing out future research needs. Table 2 presents an overview of characteristics of corporate sustainability practices through the lens of ISO 26000. To present a general view on the SR reporting practices within the case companies, we first discuss the information contents of the seven core topic themes by each organization, then build a brief comparison between the case companies on their SR practices.

Fortum is an international energy company providing customers with energy solutions that according to company strategy improve present and future life, and deliver excellent shareholder value. The company operates mainly in Nordic and the Baltic countries, Russia, Poland and India. In addition to production of heat and electricity, Fortum has recently invested in production of fast pyrolysis oil made of wood-based raw materials (e.g., forest residues, wood chips and sawdust) integrated with existing combined heat and power production and an urban district heating network.

Table 1 Background information of Finnish case companies from operating in Nordic forest and energy industries

	Fortum	Metsä group	Gasum	Koskisen
Field of industry	Energy	Forest	Energy	Forest
Turnover	3382	5016	915	247
Operating profit	808	542	126	4
Employees (#)	7835	9600	310	1057
Main product categories	Electricity, heating and cooling, energy sector services, power trading	Paperboard for packaging, tissue paper, wood products	Natural gas, bio-gas, liquefied natural gas, technical services	Plywood, sawn and processed timber, birch products, chip-board, components for wood construction
Examples of strategic renewal initiatives	Integration to circular economy with acquisition of Ekokem company	Investments in large-scale bio-product mill to supply renewable energy, materials and a bundle of (niche) bioeconomy products	Acquisition of Liquefied Natural gas (LNG) business of the Norwegian Skangas in 2014	Implementation of a project to take advantage of circular economy approach in production processes

Table 2 ISO 26000 social responsibility guideline core topics covered in case companies' communication

	Large-scale companies		SME companies	
Core topic themes	Fortum	Metsä group	Gasum	Koskisen
Governance	Commitments to UN declarations related to human and political rights, and climate, GC, ILO, International Chamber of Commerce's anti-bribery and anti-corruption guidelines Sustainability management strategy driven	Commitments to WBCSD, global compact, ILO; sustainable development goals mentioned; Diversity management; investment based strategic integration into circular economy and	Commitment to GC Participation in CIF, FEI, WEC, EC Sustainability and responsibility part of the new strategy	Commitment to the code of conduct to long-term, responsible and ethical business by taking into account personnel, environmental, financial and stakeholder needs. Core focus areas: meaningful work, a healthy environment and fair partnerships
Environment	Energy and resource efficiency, decarbonization, circular economy, reduction of environmental impacts, climate benign energy production and systems	Sustainable forest management; material and energy efficiency, emissions control and water use; renewable energy	Carbon-neutral future and innovations, efficiency and environmental impacts of the supply chain	Reduction of life-cycle impacts of production and products to soil, air and water; PEFC and FSC chain-of-custody certificates for wood
Human rights	In accordance with the UN guiding principles on business and human rights, ILO	Commitments to GC, ILO Convention	Not mentioned	Commitment to UN universal declaration of human rights. Member of FIBS Corporate Responsibility Network
Fair operating practices	International Chamber of Commerce's anti-bribery and anti-corruption guidelines Sustainable supply chain	Sustainable supply chain	Life-cycle impacts; Programs promoting sustainable development, energy efficiency and environmental protection	Circulation of wood, raw materials and energy. Systematic communication on incidents, legal actions and development
Labor issues	Operational and occupational safety	Work safety program in place	Safety and security as strategic objectives	Personnel working conditions, development of know-how,

(continued)

Table 2 (continued)

	Large-scale companies		SME companies	
				wellbeing at work, investments in health
Consumer issues	Product responsibility; guarantee-of-origin-labelled and renewable energy	Product safety; product and process innovations into new bioproducts	Openness, transparency and active dialogue; Customer satisfaction; Gas supply security	Confidentiality of contracts with customers, complying product requirements and standards, following ethical codes throughout the order-delivery chains, traceability of products and services
Community engagement and development	Participation in national and international organizations, supporting local communities	Supporting local livelihoods and society; EIA of bioproduct mill and local business ecosystem development	Supporting junior sport teams and student quild activities, project in India with UFF	Supporting local entrepreneurship, collaboration with local educational institutions
Overall assessment of guideline applicability	No direct mentioning of ISO 26000, but topics well covered, mainly due to application of GRI	No direct mentioning of ISO 26000, but topics well covered	No direct mentioning of ISO 26000, the gaps obviously due to the Nordic operation environment (no commitments to international declarations, no human rights issues)	Direct mentioning of ISO 26000; since 2015 implementation of the standard as a social sustainability framework

Sustainability is defined being an integral part of the Fortum's strategy. The company has defined sustainability focus areas in the areas of economic, social and environmental responsibility without a special focus on ISO 26000. The focus areas are based on Fortum's and its stakeholders' views of the significance of the impacts on the company and its ability to create value. The dimensions of ISO 26000 are considered very well. Consequently, the economic dimension focusses on long-term value and growth, economic benefits to stakeholders, and sustainable supply chain. Social focus area covers secure supply of heat and electricity (consumer issues in ISO 26000), customer satisfaction (consumer issues in ISO 26000), solutions for sustainable business (fair operating practices in ISO 26000), business ethics and compliance (governance, fair operating practices in ISO 26000), and operational occupational safety (labor issues ISO 26000), corporate citizenship (community engagement and development in ISO 26000), human rights (human rights in ISO

26000), and product responsibility (consumer issues in ISO 26000). The third, environmental element equals much the environment in ISO 26000 considering energy and resource efficiency, reduction of environmental impacts and climate benign energy production and systems (Fortum 2016). In 2016 Fortum acquired Ekokem, a leading Nordic circular economy company specialised in material and waste recycling, final disposal solutions, soil remediation and environmental construction. Fortum strengthened its strategy as a Nordic circular economy leader in the field of waste-to-energy solution with this acquisition.

Metsä Group is a global forest based bioeconomy group present in 30 countries, but operating mainly in Europe. At the moment, the company is building the first new generation bio-plant to be located in Finland. Abreast with pulp, the mill will produce out wood-based raw materials bio-products such as tall oil, turpentine, bio-composites and biogas, some of which have good demand potential at higher levels of Fig. 1 value pyramid.

The company addresses its current sustainability agenda under four rather general themes where ISO 26000 is not directly advocated anywhere. The first one is about offering "sustainable choices" (mainly connected to "consumer issues") , the second is about bringing raw material from forests to consumers (integrating "environment" and "consumer issues"), the third emphasizes better climate and environment (again focusing on "environment" via emission control, water foot printing and material and energy efficiency), and the fourth is about general well-being (with a clear linkage to "community engagement and development" from ISO 26000). Work safety and employer diversity programs are examples under theme "labour issues", which also fall partly under "governance". Among core ISO 26000 topics, linkage to "human rights" is via commitment to Global Compact and ILO Convention, as well via company supplier code of conducts. Promoting sustainable, renewable (forest-based) materials through production of safe, high-quality and recyclable products for the needs of bioeconomy is at the heart of the company strategy and sustainability agenda, where bioproduct mill under construction in Finland represents a flagship project and its EIA process has been a centre of company multi-stakeholder dialogue. This bioproduct mill (pulp mill based forest biorefinery) is envisioned (Metsä Group Sustainability Report 2015, p. 17) to be "a pioneer in sustainable industry with no fossil fuel CO2 emissions" and contributor with a 2% unit growth in Finnish national renewable energy strategy targets.

Gasum is a Finnish expert in natural energy gases (natural gas and biogas). It imports natural gas to Finland, upgrades biogas, and transmits and delivers these for a broad range of uses in energy production, industry, homes, and land and maritime transport. The company develops the Finnish and Nordic energy infrastructure by investing in the liquefied natural gas ("LNG") business, biogas business and transport services.

Gasum invested strongly in corporate responsibility issues in 2015. The promotion of sustainable development is considered at the core of the strategy raising safety and security as one of the key strategic objectives. Furthermore, Gasum build its roadmap as regards the transition to a carbon-neutral society by 2050. The corporate responsibility themes include carbon-neutral future and innovations, forerunner in safety and supply security, better society with the stakeholders and understanding life-cycle impacts. The majority of ISO 26000 dimensions are covered with the four elements emphasizing the environment in the first and last theme. ISO 26000 topic "government" is included in the strategic approach to corporate responsibility. The theme "forerunner in safety and supply security refers much to the topics "labor issues" and "consumer issues". Third theme, "better society with our stakeholders", refers much to community engagement and development in ISO 26000 listing (Gasum 2016). It is worth of noticing that ISO 26000 topic "human rights" is not covered in company's SR reporting, which may be a good example of the context-dependency of the responsibility documentation. As Gasum operates mainly in the Nordic area, these issues are covered comprehensively by the legislation in the region and thus it is an integral part of company's operations that these issues are being respected.

Gasum acquired a majority stake in the LNG business of the Norwegian Skangas in 2014. Skangas will continue to strengthen the position and infrastructure of LNG and the utilization of new gas solutions more extensively in Finland, Sweden and Norway replacing fossil-based crude oil. From sustainability perspective, LNG has good business potential especially in maritime transport with products having reduced sulphur emissions (see Boer et al. 2016).

Koskisen is a Finnish family-owned enterprise operating in several branches of woodworking industry by processing sawnwood and birch products together with manufacture of plywood and chipboard. In addition to focus on production processes, an important part of company's operations concerns forest management and wood procurement in its own forest holdings and collaborator private forest owners. Koskisen is also involved in renewable energy production through its sales of side-products (i.e., chips and sawdust) to the nearby power plants.

Koskisen is focused in their sustainability strategy into four main themes, which are committed family business, healthy environment, fair partnerships and meaningful work, which also form the structure of corporate responsibility report (Koskisen Sustainability Report 2015, pp. 8–15). Related to the four main themes, the implementation of CR management is based on a general code of conduct (Koskisen 2016) composing ethical instructions related to management systems (i.e., objectives to meet the international standards of ISO 14001, OHSAS 18001 and ISO 26000), human rights (i.e., following the UN human rights declaration, involvement in Finnish Business & Society FIBS agreement on business diversity at organizations in Finland), customers and suppliers, working safety, environment and wood procurement (i.e., possession of FSC and PEFC certificates) and overall implementation of corporate responsibility by leaning on ISO 26000. From the perspective of sustainability assessments, the selection of measures to reported is very narrow comprising information of only some selected issues (e.g., in the context of healthy environment the share of energy waste in relation to total amount of waste generated and in the context of meaningful work the genger distribution, worker satisfaction and sick absence). As a new opening for integrating into bioeconomy, Koskisen implemented (Koskisen Sustainability Report, p. 14) a co-operation project with Ekokem to develop circular economy strategy by finding after 10 years of development work a solution to recycle aluminum coated veneer waste.

From comparative point of view, the four companies do not illustrate any major differences in their implementation of SR. Only Koskisen is found to be directly using ISO 26000 guidance in implementing its sustainability strategy whereas other case companies with more sophisticated SR related processes and incorporation of GRI reporting for their sustainability disclosure do not seem to have need for it. On the one hand, as ISO 26000 guidance is adjusted with the GRI measurement system, following GRI guidelines also provides directly information on the fulfillment of ISO 26000 core topics whether it has been stated in organizations' strategies or not. On the second hand, since ISO 26000 does not comprise direct guidelines for selecting specific indicators to implement sustainability assessments, following ISO 26000 core topics in organizations' strategies with some unbalanced selection of few sustainability indicators is not sufficient condition to show transparently the state of SR within the organization.

Some differences that are found at least partly reflect the scale of companies, which is much larger for Metsä Group and Fortum in comparison to Gasum and Koskisen resulting in higher level of internationalization and related communications needs, for example to financers and shareholders. Gasum has among the companies most visible emphasis on national-level active customer dialogue, and with the acquisition of majority share in Skangas possibility to diversify its business with the growing potential in the use of LNG in maritime transport.

From the perspective of integration into circular aspects of bioeconomy, two different operating modes can be detected. Metsä Group has a purely investment based strategy towards integration into circular economy to diversify its value creation model whereas Fortum and Gasum have adopted acquisition based strategies using Ekokem (Fortum) and Skangas (Gasum) as stepping stones towards circular economy based business. In addition, at the same Fortum made investments in innovative fast pyrolysis production in one of its combined heat and power (CHP) production plants. However, even being at the core of bioeconomy transition, for Metsä Group the main body of attention has been with large-scale forest biorefinery under construction and the business opportunities in currently profitable pulp production. The related concerns on the sustainability and availability of biomass at competitive prices have been at the greatest focus in media, although the real possibility would seem to lie in developing an innovative local business ecosystem by integrating new players outside forest based bioeconomy. However, only time will tell after the start-up of the new bio-product mill, to what extent any positive spillover effects will emerge out to the local (circular) bioeconomy cluster. Abreast with Fortum, Metsä Group and Gasum also Koskisen has stated pursuing towards circular economy. Yet, based on the information provided by the company's internet pages, finding solutions for recycling the aluminum waste from the veneer is more related to efficient re-usage and management of waste instead of making investments on industrial processes, where those materials could be used to produce entirely new innovative and value added products in line with the profound thinking of circular economy.

6 Challenges That Case Companies Face in Promoting a Standardized View of SR

Based on case company evidence, large scale companies with more sophisticated SR related processes and incorporation of GRI reporting for their sustainability disclosure do not seem to have need for emphasizing ISO 26000 guidance. Only Koskisen was found to be using the guidance in implementing its sustainability strategy since 2015, reflecting a move towards more holistic sustainability mindset among traditional wood industry business entering the era of societal strategies for bioeconomy. In addition, for a SME company like Koskisen highly dependent on, for example, well-functioning collaboration with local forest owners and business customers in the vicinity, indicating their willingness for a more transparent SR communication may be one way of distinguishing themselves from competitiors.

In comparing results across forest and energy industries it becomes obvious that forest companies continue to be more strongly focused on environmental issues and organizational governance as key priorities for implementing their SR, while for example consumer issues and human rights receive less attention (see Mikkilä et al. 2015; Toppinen et al. 2015a, b). For all companies, community engagement practices found in place, for example, are a list of fairly traditional philanthropy oriented activities. The energy sector has met less public pressure towards its operations in comparison to large-scale forest based bioeconomy companies. This is reflected also in the implementation of SR, which is understood in the energy sector much as responsibility towards customers and employees even though the larger set of SR indicators are recognized in the GRI based disclosure.

Also strategic renewal of forest and energy companies is visible in our material. Key examples are Fortum acquiring Ekokem or Gasum becoming majority owner of Skangas. Especially the latter one is interesting. As the results of Korhonen et al. (2015) indicate, although among Nordic forest industries there have been serious concerns related to costs of tightening regulation on sulpfur emissions in maritime transport, these may be more short-run adjustments costs to be surpassed in the longer run. Substantial new business opportunities have started to materialize via adoption of clean technology in transportation and shipping, as well as via possibilities for building new strategic cross-sectoral partnerships in future development of transportation biofuels, which could be a new business area for forest biomass based producers.

To summarize the empirical findings made in this chapter, it is worth viewing ISO 26000 within the management system of a company (see Fig. 2). From there it can be observed that several benefits exist. First, ISO 26000 is directly applicable in translating sustainability principles into best practices and effective actions ("enhancing sustainability") and in building basis for defining what sustainability means for establishing a monitoring system. For the other areas of company sustainability management, the effect is of more indirect nature, especially what comes to sustainability assessment, monitoring and reporting, where other tools and auditable systems naturally already are in place. From the perspective of integrating sustainability thinking into value creation models, one can think that a guideline

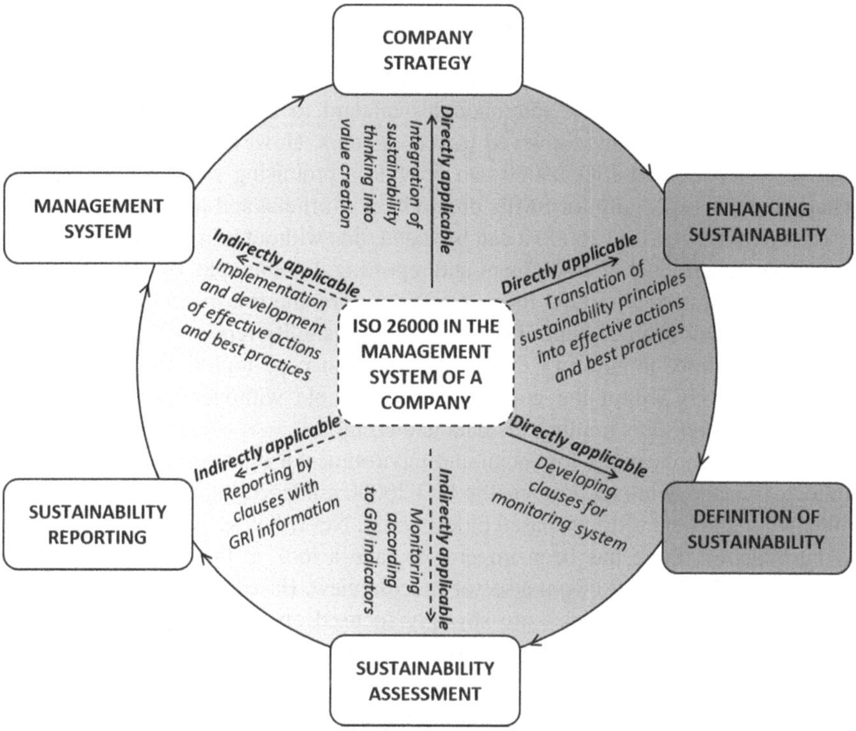

Fig. 2 Areas of direct and indirect applicability of ISO 26000 in integrating SR practices into company strategies

line ISO 26000 has a role to play when it is about building basis for introduction of the sustainability mindset into company principles.

7 Discussion and Conclusions

In comparison with earlier studies on the role of ISO 26000 for companies, for example Castka and Balzarova (2006) have argued that organizations would adopt the social responsibility agenda for strategic, altruistic or coercive reasons. As only certain organizations adopt ISO 26000, their argumentation is that organizations will most likely adopt the standard if their most salient stakeholders recognize and value ISO 26000. Hence, MNCs will seek legitimacy of their social responsibility agendas and adopt ISO 26000 if this will be the best means for dealing with it. Our analysis did not support their hypothesis, as the studied large-scale companies ignored much ISO 26000 and relied on GRI reporting.

Regarding criticism towards ISO 26000, Hemphill (2013) considered the guideline to be too broad in scope resulting in inability to capture the important environmental context of industries and sectors. Together with this, it was also found to be costly and time-consuming to implement especially for SMEs without a certifiable management system standard requiring another certifiable SR initiative to be integrated in with the international standard to allow for "legitimacy and credibility" to be publicly conveyed to stakeholders. However, the empirical findings here showed that ISO 26000 can provide a promising starting point for SR standardization especially for SMEs due to its non-official and flexible nature.

As a weakness of ISO 26000 it can be stated, that without bringing the seven core topics into actual level of assessment and reporting, for example, by integrating the management system with GRI measurement system, the transparency and sustainability communication content of the social responsibility reporting remains superficial. In addition, monitoring the development in core topics, for example, by reporting periods within the company is impossible without usage of concrete sustainability measures guiding the data gathering.

The actual implications for social and environmental improvement or corporate strategic management practices via the ISO 26000 guidance standard are largely yet unknown in the field of forest based bioeconomy. Nevertheless, few years later since its introduction, there has been impetus to take a look at the situation from the perspective of ISO 26000 from a sector specific view. Based on our findings, forest based bioeconomy companies are strongly focused on environmental issues and organizational governance as key priorities for implementing their social responsibility, while for example consumer issues and human rights still receive only moderate attention. Yet, when considering the context-dependency in SR reporting needs, this may also be a conscious and even well-grounded decision made within companies. Especially in case of Northern European companies using local raw material operating in local markets with close informal collaboration relationships with their customers, there may not be high needs for establishing reporting practices, for example, on human rights for customers.

In all, while ISO 26000 social responsibility guideline provides a relatively comprehensive framework for the implementation of corporate sustainability, it may not bring much added value to sustainability frontrunner companies with sophisticated SR related processes and incorporation of GRI reporting for sustainability disclosure. Some added value may be delivered to medium-scale companies with less sophisticated social responsibility processes. In the process of transforming traditional forest and energy industries towards bioeconomy, as such the guideline does not seem to be sufficiently detailed to incorporate forest or energy sector-specific issues and neither does it capture aspects related to circular economy processes. Yet, when combined with some existing sustainability assessment system (e.g., GRI), the core topics of ISO 26000 can support benchmarking the hot topics in companies' processes and help wrapping up social responsibility assessment information into sustainability practices that can be communicated in the society.

Acknowledgements We acknowledge financial support from Academy of Finland projects 278363 and 307480, and constructive feedback from the reviewer of this chapter.

References

Balzarova, M. A., & Castka, P. (2012). Stakeholders' influence and contribution to social standards development: The case of multiple stakeholder approach to ISO 26000 development. *Journal of Business Ethics, 111*(2), 265–279.

Boer, E., Ahdour, S., & Meerwaldt, H. (2016). *SECA assessment: Impacts of 2015 SECA marine fuel Sulphur limits: First drawings from European experiences*. Report by CEDelft. https://www.nabu.de/imperia/md/content/nabude/verkehr/nabu-seca-studie2016.pdf

BSI. (2007). *Occupational health and safety management systems (OHSAS 18001:2007)*. Retrieved August 7, 2014, from http://www.bsigroup.com/en-GB/ohsas-18001-occupational-health-and-safety

Castka, P., & Balzarova, M. A. (2006). ISO 26000 and supply chains – On the diffusion of the social responsibility standard. *International Journal of Production Economics, 111*(2), 274–286. https://doi.org/10.1016/j.ijpe.2006.10.017.

Castka, P., & Corbett, C. (2016). Adoption and diffusion of environmental and social standards: The effect of stringency, governance, and media coverage. *International Journal of Operations & Production Management, 36*(11), 1504–1529.

Ericsson, K., Huttunen, S., Nilsson, L. J., & Svenningsson, P. (2004). Bioenergy policy and market development in Finland and Sweden. *Energy Policy, 32*(15), 1707–1721. https://doi.org/10.1016/S0301-4215(03)00161-7EU.

European Union. (2009). *Regulation of the European Parliament and of the Council on the voluntary participation by organisations in a Community eco-management and audit scheme (EMAS)*. Retrieved August 8, 2014, from http://eur-lex.europa.eu/legal-content/EN/TXT/PDF/?uri=CELEX:32009R1221&from=EN

Findicator. (2017). http://www.findikaattori.fi/en/26

Forest Industries. (2017). *Industries*. Retrieved from https://www.forestindustries.fi/industry

Fortum. (2016). *Sustainability 2015, reporting content of our sustainability website as pdf.* Retrieved from http://apps.fortum.fi/sustainability/Fortum_Sustainability_2015_V2.pdf

Gasum. (2016). *Corporate responsibility report 2015*. Retrieved from http://www.gasum.com/globalassets/vuosikertomukset/gasum-corporate-responsibility-report-2015_en_final_sheets.pdf

GRI. (2014). *GRI G4 guidelines and ISO 26000:2010. How to use the GRI G4 guidelines and ISO 26000 in conjunction*. Retrieved August 8, 2014, from www.iso.org/iso/iso-gri-26000_2014-01-28.pdf

Hahn, R. (2013). ISO 26000 and the standardization of strategic management processes for sustainability and corporate social responsibility. *Business Strategy and the Environment, 22*(7), 442–455.

Hemphill, T. (2013). The ISO26000 guidance on social responsibility international standard: What are the business governance implications? *Corporate Governance, 13*(3), 305–317.

Hetemäki, L., Hoen, H., & Schwarzbauer, P. (2014). Conclusions and policy implications. In L. Hetemäki (Ed.), *Future of the European forest based sector: Structural changes towards bioeconomy, What science can tell us* (Vol. 6, pp. 95–108). Joensuu: European Forest Institute.

IEA. (2013). *Energy policies of IEA countries, Finland, 2013 review*. https://www.iea.org/publications/freepublications/publication/Finland2013_free.pdf

ISO. (2012). *7 core subjects*. International Organization for Standardization, ISO Central Secretariat. Retrieved from http://www.iso.org/iso/sr_7_core_subjects.pdf

Korhonen, J., Pätäri, S., Toppinen, A., & Tuppura, A. (2015). The role of environmental regulation for the future competitiveness of pulp and paper industry: The case of the sulfur emissions directive in Northern Europe. *Journal of Cleaner Production, 108*(Part 1), 864–872.

Koskisen Sustainability Report. (2015). Retrieved August 1, 2016, from https://www.koskisen.com/file/corporate-responsibility-report/

Koskisen. (2016). *Ethical code of conduct* (in Finnish). https://link.koskisen.com/_system/modules/digistore/InlineStream.aspx?file=RDpcQVRMQU5USVMuc2VydmljZXMtZmlsZXN5c3RlbVxkaWdpc3RvcmVcZGF0YTFcS29za2lzZW5cb3RoZXIgZG9jdW1lbnRzXEtvc2tpc2VuX2tvbnNlcm5pX2VldHRpbmVuX29oamVfZmkucGRm&d=Mjk5MjAxNg==

Marques, J. (2012). A global SR standard: Good, or too good to be true? (ISO 26000 standard for social responsibility). *The Journal for Quality and Participation, 34*(4), 29–35.

Metsä Group. (2015). *Sustainability report*. Retrieved at: http://www.metsagroup.com/en/Documents/Sustainability/Metsa-Group-Sustainability-Report-2015.pdf

Mikkilä, M., & Toppinen, A. (2008). Sustainability reporting in the world's largest pulp and paper companies. *Forest Policy and Economics, 10*(8), 500–506.

Mikkilä, M., Panapanaan, V., & Linnanen, L. (2015). Corporate responsibility in Finland – From local movements to global responsibility. In S. O. Idowu, R. Schmidpeter, & M. S. Fifka (Eds.), *Corporate social responsibility in Europe* (pp. 209–228). Cham: Springer. http://link.springer.com/chapter/10.1007%2F978-3-319-13566-3_12#page-1

Mikkilä, M., Panapaan, V., & Linnanen, L. (2016). The pursuit of responsible business: Corporate responsibility of finnish companies in their global operations (Chapter 8). In S. O. Idowu (Ed.), *Key initiatives in corporate social responsibility, CSR, sustainability, ethics & governance*. Heidelberg: Springer. https://doi.org/10.1007/978-3-319-21641-6_8

Miles, M. B., & Huberman, A. M. (1994). *Qualitative data analysis. An expanded sourcebook* (p. 338). Thousand Oaks: Sage.

Myllyviita, T., Leskinen, P., Leskinen, L. A., Lähtinen, K., Pasanen, K., Sironen, S., Kähkönen, T., Sikanen, L., & Asikainen, A. (2013). Sustainability assessment of wood-based bioenergy – a case study of eastern Finland. *Biomass and Bioenergy, 59*, 293–299.

Panwar, R., Vlosky, R., & Hansen, E. (2012). Gaining competitive advantage in the new normal. *Forest Products Journal, 62*(6), 420–428.

Parliament of Finland. (2014). *Documents of the parliament debates during the 2000s*. Retrieved from http://web.eduskunta.fi/Resource.phx/eduskunta/index.htx?lng¼fi

Pätäri, S., Tuppura, A., Toppinen, A., & Korhonen, J. (2016). Global sustainability megaforces in shaping the future of the European pulp and paper industry towards a bioeconomy. *Forest Policy and Economics, 66*, 38–46.

Rasche, A. (2010). The limits of corporate responsibility standards. *Business Ethics: A European Review, 19*(3), 280–291.

SAI. (2008). *Social accountability 8000 standard*. Reterived August 8, 2014, from http://www.sa-intl.org/_data/n_0001/resources/live/2008StdEnglishFinal.pdf, .

Stupak, I., Asikainen, A., Jonsell, M., Karltun, E., Lunnan, A., Mizaraitė, D., Pasanen, K., Pärn, H., Raulund-Rasmussen, K., Röser, D., Schroeder, M., Varnagirytė, I., Vilkriste, L., Callesen, I., Clarke, N., Gaitnieks, T., Ingerslev, M., Mandre, M., Ozolincius, R., Saarsalmi, A., Armolaitis, K., Helmisaari, H.-S., Indriksons, A., Kairiukstis, L., Katzensteiner, K., Kukkola, M., Ots, K., Ravn, H. P., & Tamminen, P. (2007). Sustainable utilisation of forest biomass for energy—Possibilities and problems: Policy, legislation, certification, and recommendations and guidelines in the Nordic, Baltic, and other European countries. *Biomass and Bioenergy, 31*, 666–684.

TEM. (2014). *Suomen biotalousstrategia (Finnish national bioeconomy strategy)*. http://www.biotalous.fi/wp-content/uploads/2015/01/Suomen_biotalousstrategia_2014.pdf

TEM. (2016). *Ministry of economic affairs and employment; energy*. Retrieved from http://tem.fi/en/energy

Toppinen, A., Virtanen, A., Mayer, A., & Tuppura, A. (2015a). Standardizing corporate responsibility via ISO 26000: Empirical insights from the forest industry. *Sustainable Development, 23* (3), 153–166.

Toppinen, A., Lähtinen, K., & Hänninen, V. (2015b). ISO 26000 in the assessment of CSR communication quality: CEO letters and social media in the global pulp and paper industry. *Social Responsibility Journal, 11*(4), 702–715.

Toppinen, A., Lähtinen, K., & Holopainen, J. (2016). On corporate responsibility (Chapter 5). In R. Panwar, E. Hansen, & R. Kozak (Eds.), *Forests, business and sustainability* (pp. 70–90). New York: Routledge.

Yin, R. K. (2003). *Case study research. Design and methods. Applied social research methods services* (Vol. 5, p. 179). London: Sage.

Anne Toppinen is Professor of forest economics and marketing at the University of Helsinki, Finland since 2008. Her speciliaty areas are economics of forest products markets, responsible business management in the global forest bioeconomy as well as in a wider scope studying sustainability transition in the context of bioeconomy. Before joining University of Helsinki she has worked in several research and research management positions including European Forest Institute, Finnish Forest Research Institute, Oregon State University and Academy of Finland. She has authored about 130 scientific journal articles and peer reviewed book chapters and well over 200 other publications, as well as supervised about dozen doctoral theses. She is a member of Scientific Advisory Board of European Forest Institute and a member in two large-scale private research granting foundations in Finland. She also acts as an associate editor in the area of economics to Forest Science, Journal of Forest Economics, Current Forestry Reports and Bio Products Business.

Mirja Mikkilä, D.Sc. (Arg. & For.), works as Associate Professor in the field of Forest-based bioeconomy and circular economy at Lappeenranta University of Technology. She is also Adjunct Professor of Forest economics and marketing of forest products, at University of Helsinki. Her fields of expertise include corporate responsibility and sustainability, environmental management and economics, natural resource economics, and sustainable transition. Her recent research interests focus on the sustainable transition on the interphases of forest-based bioeconomy and energy, food chain and textile sectors. She has published over 40 peer-reviewed journal articles, book chapters and conference proceedings. She has also over 15 years working experience within natural resource consulting and operations of pulp and paper industry on four continents, Africa, Asia, Europe and Latin America. The short- and long-term field assignments since the early 1990s covered tasks related to planning of rural development schemes, assessment of financial and environmental profitability of natural resource management and utilization, valuation of land use and development of operational management tools for bioeconomy business. A number of field assignments related to technical reports and guidelines, as well as newspaper articles have shared her research and practical findings beyond the academic audience, too.

Katja Lähtinen is Professor in Business Economics (University of Vaasa) and Adjunct professor in Forest industry marketing and management (University of Helsinki) with a PhD at the University of Joensuu, Finland. She is specialized on assessing the competitiveness and sustainability of the forest sector, especially in the context of wood industries and energy production. Currently she is working especially with themes related to wood construction business from the perspectives of strategic management within companies and their business ecosystems, customer needs driving the market demand and regulation affecting the business potential of wood construction. She has published about 30 articles or book chapters in international peer-reviewed publications. In addition, she participates actively in popularization of science, e.g., via newspaper interviews and collaboration with companies and other stakeholder groups.

Anticipating and Assessing Corporate Social Responsibility Within ISO 26000 Implementation: The Experience of Camst Cooperative (Italy)

Mara Del Baldo and Selena Aureli

1 Introduction

ISO 26000 is an ISO International Standard[1] that provides guidelines rather than requirements for social responsibility (SR). As a matter of fact, ISO 26000 is meant to extend the implementation of SR by developing an international consensus on the meaning of SR and the SR issues that organisations need to address. Moreover, it offers a practical framework for executives in charge of Corporate Social Responsibility (CSR), who face the challenge of responding to stakeholders in an effective manner (O'Riordan and Fairbrass 2008).

Nevertheless, "it is not obvious that stakeholders are at all convinced that the two key principles of responsibility and accountability espoused by the standard have been much advanced in practice" (Gray et al. 2014, p. 114). The problem is that organisations need guidance but they are at various stages of understanding and integrating social responsibility into their processes (Walker and Beranek 2015; Walker and Schmidpeter 2015) and sometimes have difficulties in practical implementation (Moratis and Cochius 2011).

Please note that this paper represents the work of a common research project. However, Del Baldo Mara wrote Sects. 2.1, 4 and 5 while Aureli Selena wrote Sects. 1, 2.2, 3 and 6.

[1]ISO is a network with a membership of 163 national standards bodies. ISO's expertise is in developing harmonised international agreements based on double levels of consensus among the principal categories of stakeholders and among countries.

M. Del Baldo (✉)
University of Urbino Carlo Bo, Urbino, Italy
e-mail: mara.delbaldo@uniurb.it

S. Aureli
Bologna University, Bologna, Italy
e-mail: selena.aureli@unibo.it

S. O. Idowu et al. (eds.), *ISO 26000 - A Standardized View on Corporate Social Responsibility*, CSR, Sustainability, Ethics & Governance,
https://doi.org/10.1007/978-3-319-92651-3_8

To solve this problem, in April 2016, the *Guidance to the application of UNI ISO 26000* (UNI/Pdr 18: 2016) was released. This is the so called 'practice of reference' dedicated to UNI ISO 26000, which aims to promote the effective implementation of ISO's principles and put them into practice. The guidance is composed of different parts, which help organisations to implement an effective SR approach and it contains several practical tools, such us the check list for the governance assessment (see guidance's Appendix B) and useful examples to support the materiality process (i.e., the questionnaire for the materiality analysis is provided in the guidance's Appendix C). It promotes a holistic and synergistic approach to SR while putting governance at the centre.[2] Moreover, the practice of reference is intended for use by organisations beginning to address social responsibility, as well as those more experienced with its implementation.

Driven by the objective to contribute to the quest for additional insights on ISO 26000 implementation, this work attempts to provide a critical analysis of the applicability and the appropriateness of this standard in the context of an Italian cooperative, named Camst, which represents one of the largest catering service firms in Italy. The company has been chosen because it is not new to SR. It already follows the SA 8000 and accounts for its actions by publishing the company's Social Report (Camst Social Report 2015). Thus, it represents the perfect setting to observe in order to understand how a more experienced user of SR concepts like Camst introduced ISO 26000 and if its adoption has led to an improvement in existing practices and a greater integration of SR into the organisation and the business model (Bocken et al. 2014; Schaltegger et al. 2012; Abdelkafi and Täusher 2015). In addition, the case study selection has been driven by the will to investigate 'peculiar' settings, different from large multinationals, like that of cooperatives, where implementation issues may happen.

The analysis is oriented towards identifying the function associated to the instrument (e.g., a tool used to increase transparency and disclosure, to improve governance or to engage stakeholders). Highlighting the role of ISO 26000 in reframing and ameliorating the CSR strategic approach of the company under investigation. In addition, it aims to identify possible gaps between the organization's current practices and the recommendations of ISO guidelines and it is used to understand whether adaptations are needed in light of specific circumstances like the fact that the company has a specific nature, i.e. it is a cooperative owned by its workers, and deals with food processing, a business sector that raises several sustainability issues regarding food waste, quality of the product, health and safety of the workers, excessive energy consumption for its distribution, etc.

The chapter is organised into two main parts: after a brief literature review aimed at tracing the theoretical framework, it presents the case-study and discusses the findings. The work provides both scientific and managerial implications. On the one

---

[2]As indicated by the ISO guidelines (ISO 26000:2010: 20), the thematic reflection on the nature of organisational governance is at the centre of the process that companies should undertake to recognise their SR because an effective organisational governance enables an organisation to take action on the other core subjects of SR.

hand, it contributes to further reflections on the effectiveness of ISO 26000. On the other hand, it suggests how organisations can maximise the benefits of ISO 26000 implementation.

2 Literature Review

2.1 Reasons for the Application of ISO 26000

ISO 26000 is a guideline of about 100 pages published in 2010, which developed after a global multi-organisational process involving several types of stakeholders and actors from different societal spheres. The purpose was to issue a document capable of supporting all types of organisations in the understanding of their social responsibilities, the development of relationships with stakeholders and the implementation of CSR practices. Different from GRI guidelines and SA 8000, its goal is not to provide rules for reporting and certification. Consequently, three main elements differentiate ISO 26000 from other regulations and norms of social responsibility. The first one is its "multi-stakeholder approach" as the guidelines have been drawn up based on the consensus among various types of actors. The second element refers to "globality" due to the fact that the ISO working group was made up of experts coming from all over the world, including developing countries. The latter is related to the importance attributed to social dialogue and negotiation with stakeholders, which marks the positive and central role an organization can and should play in promoting a social dialogue with its stakeholders, starting from its employees.

ISO 26000 appears to be strongly in line with the renewed concept of CSR proposed and promoted at the European level and it supports the objectives of the Europe 2020 strategy for "intelligent, sustainable and inclusive growth" (EC 2011). ISO 26000 is in the line with Global Environmental Management Initiative (Burlea Schiopoiu 2013) and emphasizes the importance of results and improvements in performance on social responsibility. Therefore, ISO 26000 provides a valuable guidance to all type of organizations, regardless of their size or location, on:

- concepts, terms and definitions related to social responsibility;
- the background, trends and characteristics of social responsibility;
- principles and practices relating to social responsibility;
- integrating, implementing and promoting a socially responsible behavior throughout the organizations and, through the policies and practices, within the sphere of influence of each organization;
- identifying and engaging with stakeholders; and communicating commitments, performance and other information related to social responsibility.

The ISO 26000 document consists of seven chapters (ISO 26000 2010). Notably, chapter 4 describes the seven principles of social responsibility: Accountability (an organisation should be accountable for its impacts on the society, the economy and the environment; Transparency (an organisation should be transparent in its decisions and activities that impact the society and environment); Ethical behaviour (an organisation should behave ethically); Respect for the rule of law (an organisation should accept that respect for the rule of law is mandatory); Respect for international norms of behaviour (an organisation should respect international norms of behaviour, while adhering to the principle of respect for the rule of law); Respect for human rights (an organisation should respect human rights and recognise both their importance and their universality); Respect for stakeholders' interests (an organisation should respect, consider and respond to the interests of its stakeholders). Chapter 6 is also very important has it indicates the seven core subjects or areas of SR in which organizations should account for their actions and compliance with the principles previously described (Table 1). All core subjects comprise a number of issues; every organization (included governmental organization) should identify which issues are relevant and significant based on its own considerations and through a dialogue with stakeholders.

The standard guides organisations through a three-stage process:

- the identification of company stakeholders and their engagement;
- the identification of how the organisation's decisions and actions impact stakeholders and the society in general (evaluation of actual and potential impacts through the analysis of seven core subjects and associated issues);
- the consideration of how to apply SR (i.e. how to integrate it in the organisation's strategy, culture and operations, how to communicate it and how to improve it).

Despite its guidance nature, ISO 26000 is considered and termed as standard for social responsibility by both the literature and the International Standard Organisation that issued it (Balzarova and Castka 2012; ISO 2010). More precisely, ISO 26000 should be considered a principle-based and guidance-based standard, which attempts to translate seven principles of SR into suggestions for implementation.

Similar to other CSR standards, guides and frameworks, ISO 26000 believes that embracing social responsibility leads to several benefits like improving company reputation, obtaining a competitive advantage, retaining talented employees as well as customers, reducing the cost of capital thanks to the increased transparency and improving relationships with suppliers, governments and other subjects (ISO 2010; Hemphill 2013). At the same time, it differs from other standards for being less oriented toward supporting organisations' external accountability and more focused on providing companies with a practical tool to measure and nurture the governance of the organisation (Hemphill 2013), being that it indicates how to achieve certain sustainability goals (Katamba et al. 2014). In other terms, it has the potential to guide companies on their path to sustainability and contribute to their strategic management processes (Hahn 2013).

Sustainability standards have been developed for three different purposes: to increase a company's transparency toward stakeholders, improve their internal governance and engage stakeholders (Gray et al. 2014). ISO 26000 seems to be more concentrated on encouraging company executive leadership to carry out a thematic reflection (the seven core subjects to consider) on its management and related results. Its goal is to embed a deeper understanding of CSR into the

Table 1 ISO 26000 Core subjects and issues

Core subjects	Issues
Organizational governance	Overview of organizational governance Principles and considerations Decision-making processes and structures
Human rights	Issue 1: Due diligence Issue 2: Human rights risk situations Issue 3: Avoidance of complicity Issue 4: Resolving grievances Issue 5: Discrimination and vulnerable groups Issue 6: Civil and political rights Issue 7: Economic, social and cultural rights Issue 8: Fundamental principles and rights at work
Labour practices	Issue 1: Employment and employment relationships Issue 2: Conditions of work and social protection Issue 3: Social dialogue Issue 4: Health and safety at work Issue 5: Human development and training in the workplace
The environment	Issue 1: Prevention of pollution Issue 2: Sustainable resource use Issue 3: Climate change mitigation and adaptation Issue 4: Protection of the environment, biodiversity and restoration of natural habitats
Fair operating practices	Issue 1: Anti-corruption Issue 2: Responsible political involvement Issue 3: Fair competition Issue 4: Promoting social responsibility in the value chain Issue 5: Respect for property rights
Consumer issues	Issue 1: Fair marketing, factual and unbiased information and fair contractual practices Issue 2: Protecting consumers' health and safety Issue 3: Sustainable consumption Issue 4: Consumer service, support, and complaint and dispute resolution Issue 5: Consumer data protection and privacy Issue 6: Access to essential services Issue 7: Education and awareness
Community involvement and development	Issue 1: Community involvement Issue 2: Education and culture Issue 3: Employment creation and skills development Issue 4: Technology development and access Issue 5: Wealth and income creation Issue 6: Health Issue 7: Social investment

Source: Our elaboration from ISO 26000 (2010)

organisation (Heidi von Weltzien Hoivik 2011). Thus, it should not be considered an unnecessary repetition of previous standards (Zinenko et al. 2015) nor a separate alternative to UNGC, GRI or other standards. These tools are complementary to each other because they have different goals and are useful in different areas of an organisation's CSR infrastructure (De Deus et al. 2014; Moratis 2016a).

Needless to say, the concrete application and the results expected from the adoption of ISO 26000 are believed to be different in the case of more experienced companies that have already implemented other sustainability frameworks or codes of conduct. These companies may use the standard as a simple 'check tool' to identify possible gaps in current practices or as a 'holistic reference' to address minor issues related to policies and practices (Hemphill 2013). On the contrary, companies that are new to SR may use it to start integrating sustainable activities into the business and obtain greater benefits.

The adaptability of the standard to different contexts claimed by ISO itself is probably one of the main reasons for the growing uptake of ISO 26000 implementation in different types of companies all over the world (ISO 2012). However, additional reasons can be mentioned: the positive image of ISO as a globally reputable and credible organisation for establishing international technical standards (Hemphill 2013); its broad-based and multi-stakeholder development process (Hahn and Weidtmann 2016), which gives the standard a high international consensus on its definition of SR, a considerable degree of legitimacy (Balzarova and Castka 2012) and the potential to become an important guide for firms worldwide (Mueckenberger and Jastram 2010; European Commission 2011); its origin based on ISO participants' best practices, which makes it look as feasible and reliable for management teams interested in integrating social responsibility principles into enterprise operations.

2.2 Issues in the Standard's Implementation

Despite the cited benefits that provide reasons for the adoption of ISO 26000, there are several obstacles that may hinder its diffusion.

Firstly, some authors have highlighted that the standard is not adaptable to every business regardless of its claim of being "useful to all types of organisations in the private, public and non-profit sectors, whether large or small, and whether operating in developed or developing countries" (ISO 2010, p. iv). Industry representatives played a major role in the discussion process of the definition of the standard, while other stakeholders didn't have the possibility to participate in international meetings, therefore, their voices have not been included (Schwartz and Tilling 2009; Balzarova and Castka 2012; Boström and Hallström 2013). In addition, ISO has a history of involving large multinationals in former processes devoted to the creation of management system standards like ISO 9000 and ISO14000 (Balzarova and Castka 2012). All of these aspects suggest that ISO 26000 has been designed to better suit large corporations instead of small and medium-sized enterprises (SMEs).

Perera (2008) and Hemphill (2013) reinforced this matter by demonstrating that ISO 26000 has several limitations that can make its application problematic in SMEs. One of the major obstacles is represented by the great amount of time and resources required for the standard implementation, which is not offset by the benefit of obtaining a CSR certification and communicating it to the public. Thus, SMEs may prefer adopting GRI guidelines, which can be certified by an independent third party and increase company reputation.

Secondly, the standard is considered too broad in scope and poorly detailed to be useful in specific industries and sectors (Hemphill 2013) like the forest sector examined by Toppinen et al. (2015a, b). This aspect also raises doubts about its appropriateness and applicability in specific contexts like those of cooperatives.

A third critique refers to its limited benefits for companies with a long history of sustainable practices. Some authors believe that the standard is more useful for beginners, therefore, it does not bring much added value to sustainability frontrunners (Hahn 2013; Toppinen et al. 2015a, b).

Moreover, since it is a guideline and not a standardised management system, its adoption cannot be certified (Moratis 2016b), making ISO 26000 unsuitable for contractual or regulatory use. This aspect hinders ISO 26000 to become a proper instrument to signal CSR commitments and performance of firms, possibly compromising the standard's further adoption (Moratis 2016b).

Another issue refers to the absence of specific requirements for managers to identify acceptable solutions for the negative impacts of company decisions and activities on the society and the environment (Johnston 2011). The standard helps companies to learn about the 'externalities' caused by their operations and what social responsibility entails in a particular context, but it does not require them to indicate how they shall bridge the gap between identifying social and environmental impacts and making sustainable decisions (Johnston 2011).

With all these issues in mind, the present work aims to investigate ISO 26000 implementation and its current usage to shed light on the possible gaps between actual company practices and stated recommendations from the ISO guidelines. The case study analysis allows comprehending how the implementation process has evolved and which factors favour or hinder the realisation of ISO 26000's expected benefits.

3 Research Method

Because the objective of this research is to provide a critical analysis of the implementation of ISO 26000 by observing peculiar contexts like that of cooperatives, where implementation issues may arise, researchers decided to recur to case study analysis, which allows to better understand how processes develop in real life. Accordingly, the empirical study has been developed using a qualitative approach (Patton 2002; Bailey 2007; Smith 2011) and a case-study method (Yin 2009; Eisenhardt 1989; Eisenhardt and Graebner 2007; Naumes and Naumes 2006).

Within the emerging field of research on ISO 26000 there are several studies that recurred to the case study methodology (von Weltzien Hoivik 2011; Katamba et al. 2014; Ranängen et al. 2014) with the aim to understand how ISO 26000 is conceived and used within enterprises and to analyse if its application can improve a company's CSR practices. On the contrary, quantitative-based studies on ISO 26000 use the standard as a framework to identify and assess company efforts in social responsibility practices, such as stakeholder identification or stakeholder engagement, which are either declared in reports or reported by interviewees (Zompras and Siakas 2015; Valmohammadi 2014). In other terms, quantitative studies do not help us to understand why and how SR processes are implemented; they merely describe and measure the overlap between ISO 26000 recommendations and the actual practices of a company (Toppinen et al. 2015a, b), trying to verify if ISO 26000's implementation is linked to better financial, social and environmental performance (Habidin et al. 2014; Peršič et al. 2016).

Since case study analysis outperforms when researchers use a combination of methods with which they collect research materials (Bailey 2007), the present work recurs to both interviews, a questionnaire and document analysis.

In-depth semi-structured interviews were conducted with the company's interlocutors involved in varying CSR activities and strategies. The interviews (managed by telephone) were preceded by e-mail correspondence and the submission of a questionnaire. Three key subjects were interviewed: two of them are Camst managers, while the third one is an external independent assessor. One interviewee is responsible for standards, regulations and social responsibility standards/systems; she takes care of CSR initiatives regarding different levels and aspects (e.g., the implementation of certified systems and relationships with shareholders). She has been managing the ISO 26000 and SA 8000 implementation for more than a year. The other interviewee is a Camst Board member and he is the chair of the 'Social Relations Department'. As part of this function, the interviewee handles relations with shareholders, presides over Camst's Social Report and deals with external subjects such as independent assessors, assurance service providers and consultants for CSR (for example Impronta Etica). The external independent assessor is the Social Responsibility Sector Manager for the Italian subsidiary Bureau Veritas (Bureau Veritas is a world leader in the assessment, inspection, testing and certification of companies for quality, health & safety, environmental and social aspects). Each interview lasted about 1 hour and a half. All interviews were transcribed, validated and then analysed in order to identify relevant aspects.

The interviews were supplemented with the document analysis of data downloaded from Camst's corporate website and its Social Report as a means to better understand how the selected company managed the implementation of ISO 26000. Researchers also used the documents issued by ISO (ISO 2010) and public reports published in collaboration with other organisation (e.g., UNI-Soldalitas 2014). In total, the analysis lasted 6 months, from February 2016 to July 2016.

The questionnaire and the interviews aimed to inquire the following topics:

- Camst's reasons for implementing ISO 26000;
- how Camst implemented the standard (i.e. if it performed a total or partial implementation of the ISO 26000 guidelines) and which principles have (not) been implemented or are still in progress;
- the benefits achieved with ISO 26000 implementation;
- the factors that favoured the application of ISO 26000;
- and the obstacles that hindered, slowed down or prevented its application.

Collected information has also been used to understand Camst's elements of the trinomial mission-corporate governance-accountability[3] model (Matacena 2010), which are the keys of interpretation for understanding every business reality and are tied by reciprocal relations and coordination. Camst has been selected because of its long lasting involvement in SR. In addition, the researchers choose Camst because of: (1) the relevance of cooperatives with respect to the Italian socio-economic context; (2) the greater diffusion of ISO 26000 among medium-sized and large corporations—like Camst—as revealed by a recent survey (UNI and Sodalitas 2014); (3) Camst's cooperative nature, in which its capital is owned by its own workers/employees, which may offer a peculiar context for analysing its ISO 26000's implementation in contrast to the studied setting of multinational corporations and (4) Camst's business regarding food processing and distribution, which represents a key stage in the food supply chain capable of making a significant contribution to climate changing greenhouse gas emissions, generating other major environmental impacts and influencing society's future survival.

4 Case Study Presentation: CAMST's Experience on ISO 26000

Company Profile

Camst is a production and labor cooperative, specialised in all sectors of catering from school catering to the provision of canteen meals, and ranked amongst the leading companies of this type of business in Italy. It belongs to the segment of collective/organised catering, whose turnover amounts to 4782 million euros (Databank 2015 as cited in Bilancio Sociale Camst, 2015, p. 102).[4] Its management and administrative headquarter is located in Villanova di Castenaso (Bologna). With several local production and management sites (about 2000), it mainly operates in the central and northern part of the country. The Camst Group's market share amounted to 8.2% in 2015, ranking it second after the Elior Group. In the sub-segment of commercial restaurants, it ranks 9th out of 11 company groups (with McDonald in the lead). Camst's performance data in December 2015 are provided in Table 2.

Camst was founded in Bologna in 1945 by 16 employees working in the hotel and canteen services, who decided to join their forces to create a cooperative made up of waiters, assistant cooks, bartenders and other restaurant workers in a historical period characterised by economic difficulties and a high rate of unemployment. As a cooperative, Camst adheres to the *National League of Italian Cooperatives* (Legacoop) and indirectly to the *Italian Cooperative Alliance*. Overall, the impact

[3]The mission is intended as a synthesis of a company's aims; the governance as the system of government and power of an organisation and the accountability is conceived as the company's responsibility to take into account and give account to stakeholders for all activities and choices.

[4]www.databank.it

Table 2 Camst Group's main financial figures

Total operating revenues	485,218,000 euros
Gross margin	43,812,000 euros
Net earnings	4,084,000 euros
Total turnover	627,410,000 euros
Share capital	21,522,433 euros
Capital owned by working shareholders (shareholders that are employed by Camst)	10,981,959 euros
Capital owned by financing shareholders (shareholders who only provide funds)	10,540,474 euros

Source: Camst's annual report

of the cooperation on the Italian GDP is about 8% (with the exclusion of cooperative banks) and represents 34% of the distribution and consumption retail, with 35 billion euros 'Made in Italy' agri-food production.

Today, the Camst Group operates through more than 20 companies. Most of them are fully owned or affiliated. They provide both catering and integrated services to individuals, companies and public bodies. Camst has also created some joint public-private companies, based on the collaboration with public entities, in order to offer complementary activities. It employs about 13,000 workers, of which 1350 are chefs, producing 115 million meals a year. Camst was the first company in Italy to enter the school canteen market. In commercial catering, Camst is widespread throughout towns and cities, operating with different brands. It is also among the leading companies in the catering sector for companies, producing 25,5 million meals a year and managing both internal company canteens and delivered meals. Lastly, Camst works for about 180 public and private health care structures, such as hospitals, private clinics and rest homes and sells modified atmosphere meals for cafeterias, hotels and vending machines.

Mission

Camst has based its activity on three building principles, which are part of its mission (1994) and its cooperative statute: (1) "voluntariness and open door": anyone can become a cooperative shareholder; (2) "one person, one vote": all members are equal, and (3) "mutuality and solidarity": everyone offers their own contribution and benefits from all of the advantages. The mission was updated in 1999 with the inclusion of an explicit reference to Environmental Protection (see Table 3) and in 2006, the four articles of the mission became an integral part of the Policy for Social Responsibility.

Governance

Camst is a cooperative with limited liability, which envisages two categories of shareholders or cooperative members: working shareholders and financing shareholders. Only the first shareholders are employed by the cooperative, thus being workers and shareholders at the same time. However, even non-member workers are employed. Shareholders are widespread in almost all Italian regions.

Table 3 Camst's cooperative mission

Article 1	Article 2	Article 3	Article 4
Defend and develop the employment of members. Improve their professional qualification and their social and working conditions. **Ensure respect of cooperative members/shareholders' health, their moral integrity, their cultural and civil aspirations** and their participation in the cooperative	Generate revenues and allocate profits in a way that maximises the development of company investment and **recognises everyone's individual contribution**	Satisfy the needs of customers, while **ensuring the product and service quality and safeguarding the environment**. Maximise competitiveness and efficiency, innovating the enterprise. Pursue and maintain the leadership within the catering sector in Italy	**Collaborate** with other cooperatives on the basis of **solidarity** and entrepreneurial rigour in order to strengthen market cooperation and favour the development the cooperative's **social role, respecting the traditions and the local cultural heritage**

Source: Camst's Social Report (2015)

The cooperative's governance system is based on the Camst's Statute, the various internal regulations approved by company shareholders and the Code of Ethics, proposed by the Board and approved by the Assembly. The cooperative organs include: the Assembly, the Board of Directors (formed by 12 members, plus the President and Vice President) and the Board of Auditors. The Board of Directors mainly has a representative and political function, while strategic and operational decisions are made by the Direction, which groups executive directors. Thus, there is a sort of dual institutional model. The Direction (see Table 4) includes the General Director, the Operations Director, the Directors of Central Services and the Directors of Territorial Divisions. The organ responsible for managing Camst's relationships with its stakeholders is the Social Affairs Directorate.

Accountability

Camst gives accounts to shareholders for its activities through the Social Report. The drafting process of the Social Report is managed by the President with the support of other corporate functions. In addition, Camst has an established policy regarding corporate social responsibility, which has been formalised in a document that incorporates the four articles of the mission. The document is signed by top management and is made available to all employees and stakeholders through the Social Report. On the basis of this policy, Camst defines specific objectives subjected to a periodic review.

In 2006, Camst adopted the SA 8000 standard (the first international standard aimed to measure the ethical and social responsibility degree of an organisation). Following the perspective of continuous improvement, in 2015, Camst initiated a process of integration of its social responsibility approach with the ISO 26000 guidelines. This has led to redefining a section of the Social Report, in line with the development of this process as specified in the following section.

Table 4 Camst's government organs

Number and type of members	Women	Men	Total
Board of directors	7	7	14
Direction	1	19	20
Average members' age (years)			
Board	50,0	52,3	51,1
Direction	61,0	51,5	52,0
Average seniority (years)			
Board	22,7	21,7	22,2
Direction	34,0	15,8	16,8

Source: Camst's Social Report (2015, p. 27)

5 Managing Camst's Sustainability with the ISO 26000

5.1 Purpose of the ISO 26000 Implementation

In line with its cooperative nature, Camst has always dealt with social responsibility, paying particular attention to the respect of human rights, workers' needs and environmental issues. The company declares: "*This is all part of our daily way of working*" (Camst's Social Report 2015, p. 29). Its milestones in SR are represented by introduction of SA 8000 in 2006 and the adoption of ISO 26000 in 2015, after years of "experience in certifications" related to product and socio-environmental aspects.

Since food safety represents a fundamental aspect in regard to the industry in which the group operates and is a prerequisite for the participation in public tenders, Camst originally began to set up processes for guaranteeing the proper preparation of its products in order to then implement additional activities devoted to allowing the company to manage and control all corporate activities in terms of quality, food safety, traceability, environmental protection, health and safety at work and SR. All of these activities are now combined into an Integrated Management System, inspired by ISO 26000, that includes the following:

- Food Quality, Safety and Traceability Management System (certified by UNI EN ISO 9001:2008) related to the product offered;
- Environmental Management System (certified by UNI EN ISO 14001:2004);
- Safety Management System (certified by OHSAS 18001:2007) related to work conditions;
- Company Social Responsibility Management System (certified by SA 8000).

Independent certification authorities regularly verify the conformity of the cited individual Management Systems according to mandatory or voluntary regulations. The ISO 26000 integrates all aspects of the individual systems since it operates as a "global key for reading all SR issues and tools for sustainability implemented" (Strasserra, 21 July 2016).

The Company also acts in accordance with the principles contained in the UN Declaration of Human Rights and the ILO Conventions and Recommendations. These certifications demonstrate Camst's sustainable orientation, which results in the commitment to customers, the protection of the environment and health and safety in the workplace. Coherently, sustainability indicators include economic, social and environmental performance indicators.

The decision to adopt the ISO 26000 guidelines took place in 2015 based on the company's willingness to integrate SR objectives into the corporate strategy. Camst was aware of the centrality of ethics in every business activity and wanted to solve its communication problem with both its internal and external stakeholders due to its multi-site company nature. In fact, having offices spread throughout Italy, SR was affected by different and sometimes inconsistent interpretations. "The decision to pursue the guidelines of ISO 26000 is a choice that comes naturally to us because the principles of social responsibility are part of our DNA" (Bertocchi, 21 July 2016). Thus, the primary functions attributed to ISO 26000 are to "enhance corporate SR creating efficiency in existing processes" (Camst's Social Report 2015, p. 29) and suggest a more coherent, structured and systematic communication and reporting activity addressed to all stakeholders.

The implementation of ISO 26000 was promoted by the staff function "Direction of Quality Control" and sponsored by the Chairmen of the Direction.

5.2 Implementation

A key aspect of the analysis refers to interviewees' evaluation of Camst's level of application of ISO 26000. After 1 year, the company examined defined the areas or core subjects related to SR but it has not yet implemented the materiality process in accordance with the practical guidelines recently issued (UNI/PdR 18:2016). Notably, Camst has recognised its SR by identifying the different issues raised by the impacts of its decisions and actions, but it is trying to improve the identification of key issues and relevant stakeholders by introducing a materiality analysis. Moreover, it still has to involve internal and external stakeholders (i.e., activate a process of stakeholder engagement) through which they can define how to apply SR within the organisation, report CSR actions and strategies and revise them with the logic of continuous improvement.

Thus, with reference to the first two phases of ISO 26000 implementation (i.e., the identification of company stakeholders and the identification of how the organisation affects them), Camst has identified its key interlocutors (Fig. 1).

Subsequently, Camst has assessed each core subject of ISO 26000 recurring to a procedure of analysis based on four perspectives (Table 5). The four perspectives helped managers to identify the megatrends and the material issues within the CSR issues, which the top managements assigned a priority (Table 6). Camst's reported issues do not strictly repeat the list provided by ISO guidelines being that "*the core*

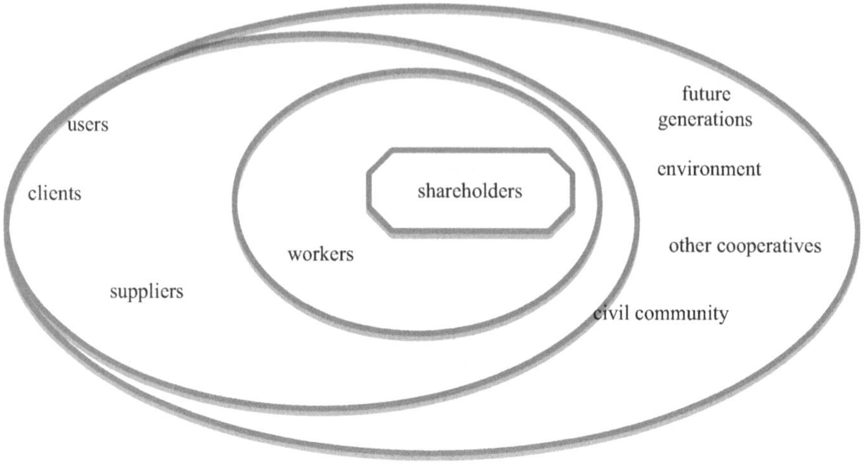

Fig. 1 Key stakeholders. Source: our elaboration from CAMST's Social Report (2015)

subjects and their respective issues can be described or categorised in various ways" (ISO 26000: 2010, p. 15).

In addition, the four perspectives helped managers to make an internal self-assessment on SR that is useful to prepare an action plan. These perspectives can be considered as a grid of evaluation that has been applied to the core subjects suggested by ISO 26000 to better understand what the company is doing with reference to SR, which programs have been implemented and how the company can improve. As suggested by ISO guidelines, issues should not be merely listed; each issue should be described together with its related expectations and actions.

The self-assessment method mentioned above is managed by the "Quality control Department" staff, which also involves managers from other departments (e.g., those who deal with public tenders and interacts with the stakeholder—"Public Administration"). The analysis was performed through individual meetings and interviews, aimed at understanding managers' perceptions and how certain practices are *de facto* conducted. The analysis involved the definition of scores for each core subject. These scores were assigned based on the managers' responses from a given set of questions. The scores helped measure the degree of Camst's SR maturity as suggested by the practice of reference issued in 2016.

Another important process implemented by Camst refers to the introduction of a gap analysis of its governance process and reporting activities dealing with CSR, aimed to highlight the distance between current company practices and what is suggested by ISO guidelines and the practice of reference issued in 2016, in order to identify possible areas of improvement. A specific work group (which also includes the two managers interviewed) was set up in order to develop the gap analysis with the external independent assessor Bureau Veritas and then to critically identify the areas of improvement.

Table 5 Perspectives guiding the assessment of relevant issues

Perspectives	Object of investigation
Integration	Camst's ability to extensively implement the seven ISO principles in the various company departments
Stakeholders engagement	Camst's ability to make dialogue with stakeholders in order to grasp their expectations and/or share projects of common interest
Communication and transparency	Camst's ability to report and communicate objectives, projects and results to both internal and external stakeholders in an accessible way
Results and progress	Camst's ability to operate in a systematic way and achieve the desired results with the logic of continuous improvement

Source: Camst's Social Report (2015, p. 30)

Table 6 List of Camst's priorities

Areas	Issues
Economics and governance	• Creation of economic value • Supply chain management • Fight against corruption • Risks prevention
People	• Increased employment • People growth/development • Quality relationships with employees (which includes the daily work relationship and compensation issues) • Health and safety at work • Labour relations • Value of diversity • Development and community involvement • Protection of human rights • Privacy
Products	• Quality and product safety • Appearance of nutritional products • Accessibility and quality of services • Information and product communication
Environment	• Waste management • Energy efficiency and emissions • Managing water resources • Fuel management • Protection of biodiversity (biological, zero km, GMO free, etc.)

Source: Camst's Social Report (2015, p. 30)

The gap analysis can be conceived as a benchmarking activity that should help the company to review and improve its own practices. Thus, it mainly refers to the third stage of ISO 26000's implementation process.

Lastly, it is interesting to note that Camst asked Bureau Veritas to publicly assess what it had done so far with reference to its organisational governance. This external assessment led to the publication of a statement, prepared by Bureau Veritas and addressed to stakeholders, that demonstrated Camst's involvement in SR, probably increasing its credibility. The statement declares that Camst has a system of policies, procedures, responsibilities and controls, which allows the organisation to take into account the expectations of stakeholders and to translate planned objectives into actual results. At the same time, Bureau Veritas released an internal report, reserved for management, to help them understand the maturity of Camst's approach to SR.

5.3 Emerging Benefits and Difficulties in Implementing ISO 26000

With reference to expected benefits and results achieved, respondents have stated that there haven't been significant changes in how the company shapes and implements SR actions, which are directly attributable to the introduction of ISO 26000. According to them, at least 3 years are needed to appreciate the real effects and benefits of its application. However, other important results can be listed.

Firstly, Camst ameliorated its self-assessment method in order to identify relevant issues related to social responsibility. "*Until now Camst had its own method of self-assessment devoted to the evaluation of the company's commitment in reference to corporate social responsibility. But now, with the support of ISO 26000, this practice of self-evaluation has been greatly improved and systematised*" (Strasserra, Bureau Veritas, July 21, 2016).

Self-assessment is crucial for measuring the process of stakeholders' engagement, identifying relevant SR issues and for improving of both internal and external engagement using different methods (i.e., workshops, forum and round tables), starting with listening, consultation, involvement of stakeholders and moving towards partnership, empowerment and engagement of the stakeholders. The latest phase named engagement includes the delegation of the stakeholders of key strategic issues and presupposes the full commitment of top management and the involvement of the entire organisation. "*The introduction of the self-assessment tool in Camst should be considered particularly relevant since it underlines the pioneering approach of this company and its willingness to improve the SR strategy and render it effective*" (Strasserra, Bureau Veritas, July 21, 2016).

Secondly, in accordance with the practical guidelines (UNI/PdR 16: 2016), Camst also learned how to perform a gap-analysis aimed at verifying the distance between the actual and the desired level of SR performance concerning governance and reporting. This tool has the benefit of generating the company's awareness of the fulfilment of its goals (if formulated) and improving managers' decision-making process, because it suggests measures to be taken. In other words, the gap analysis highlights changes compared to set targets or, as in Camst's case, compared to general objectives defined by the standard and then identifies areas of further improvement.

Finally, the project implemented also contributes to the improvement of the company's reputation and legitimacy. In June 2015, Camst submitted its application for ISO 26000 to the external evaluation and assessment of Bureau Veritas, which released a declaration addressed to stakeholders and a private report to management about the maturity of company's approach to social responsibility. This is not a certification; however, as stated in the company's Social Report, "*the declaration of Bureau Veritas strengthens the credibility of our involvement in SR*" (Camst's Social Report 2015, p. 29) and has a positive impact on the stakeholders' perceptions.

Difficulties in implementation were limited because the ISO project has been authentically promoted by top management and supported by internal engagement. "*If the project did not have the 'sponsorship' of the Chairmen, it would not have been possible to overcome the difficulties encountered in the first phase*" (Minarelli & Bertocchi, Camst Officer, July 21, 2016). Difficulties encountered so far are related to the formalisation of the materiality process and the internal and external communication of SR. In addition, an important aspect has been highlighted. Interviewees said that the implementation of ISO 26000 requires companies to raise SR to the strategic level and abandon the 'certification approach'. As one respondent claimed: "When you introduce ISO 26000 guidelines, *you do not work to get short-term result, clearly stated and easy to gain such as a specific change in a process which allows you to obtain a certification*" (Minarelli, July 21, 2016).

6 Discussion and Conclusions

Drawing from the empirical analysis, this section reviews the main findings related to ISO 26000's implementation and the future goals relative to Camst's journey within SR.

Results indicate that a more experienced company in SR, like Camst, may obtain relevant benefits from the introduction of the ISO 26000 guidelines. Its adoption has involved the introduction of new assessment methods and a greater awareness of how the company is performing in terms of SR. The implementation of ISO 26000 has also activated a participatory dialogue process (von Weltzien Hoivik 2011), which mainly involves internal stakeholders, leading to a deeper understanding of CSR within the organisation. Another important aspect refers to ISO 26000's contribution to the strategic management of SR because in Camst the 'task of ISO 26000' is to render SR an integral part of company life by creating efficiencies in existing SR processes and among them (this company already adopted a SR processes before the introduction of the new standard). In addition, the adoption of ISO 26000 enhanced Camst's credibility and reputation, since the company collaborated with a well-known independent institution like Bureau Veritas to implement some tools and to get a public evaluation of what they have been doing in terms of SR.

Cited benefits suggest that the ISO 26000 guidelines may be implemented for a specific reason (as a governance tool in the case of Camst and to communicate with stakeholders), but its application can generate additional outcomes that were not expected such as the increase in the company's credibility obtained through its collaboration with Bureau Veritas.

With reference to the implementation process, results indicate that difficulties in the adoption of the standard are always to be included. In the case study, the main difficulties were tied to the involvement of external stakeholders. "*While it appears that the self-assessment process is successful in engaging internal stakeholders and communicating to them the relevance of sustainability issues for the company, there are still gaps to fill in relation to the external ones*" (Minarelli, 21 July 2016). As stated during the interviews, Camst realised that the application of ISO 26000 does not happen overnight. It requires a different approach compared to those previously adopted in order to obtain certifications, and this aspect may generate organisational difficulties if not properly addressed. "*The managers involved in this process have understood that ISO 26000 is a different system, in which the main objective (and benefit) is to stimulate a process of continuous improvement and not to obtain a certification*" (Bertocchi, 21 July 2016).

Conflicts with other existing management systems related to sustainability issues were not mentioned. ISO 26000 perfectly integrates with all other SR management systems and related certifications. Actually, ISO 26000 is "transversal" to other certifications. "*This aspect has been a key driver for engaging several managers and different organisational departments and areas on the common theme of social responsibility, especially in the self-assessment phase. ISO 26000 leveraged the collaboration of departments and people that probably would've never been directly involved in the case of a specific certification. This benefit should not be underestimated, especially if we consider that, in Camst, similar experiences of "synergies activation" have not been performed in the past*" (Bertocchi, 21 July 2016).

No difficulties in performing the gap analysis were mentioned. The reference of practice issued in 2016 indicates that when companies try to highlight variances in expectations they usually find difficulties in several aspects, related to the identification of key performance indicators (in terms of quantitative and quantifiable measures), the lack of "resilience" in improving the process, the monitoring of key performance indicators (due to the lack of people to whom assign a specific responsibility of control), the lack of partnership with stakeholders and the redundancy of data and information. But this was not the case.

Results achieved in ISO 26000 implementation may be partially attributed to the participation of an external, qualified third party assessor such as Bureau Veritas. The institution helped Camst with its self-assessment, bringing added value to the process and suggesting procedures, and it also helped Camst to escape from the logic of self-reference.

Unfortunately, some criticalities still exist, which also represent future goals to be fulfilled. An unsolved problem is the lack of a coherent and unified supervision of processes and initiatives related to SR. "*We have done several things but the initiatives taken to improve our SR remain fragmented*" (Bertocchi, Camst, 21 July 2016). In addition, Camst still has to overcome some problems in the application of the materiality principle and the formalisation of SR strategies and activities. The analysis of materiality is fundamental for the identification of key areas on which Camst should focus its attention and efforts. However, it is linked to stakeholder engagement. Thus, it requires several years in order to be designed and implemented through the involvement of company stakeholders. With reference to this, the case of Camst is not isolated. The report of Uni and Sodalitas (2014) revealed that materiality is a critical aspect for all of the companies that try to integrate CSR into their company processes. Moreover, the report indicates that there are still few companies that include external stakeholders in defining strategies and targets for sustainability.

Despite the fact that the respondents declared that it takes 3 years for the appreciation of ISO 26000's benefits, they also marked that without the stimulus of ISO 26000, the path of pursuing a common and shared vision of social responsibility could be even longer. The introduction of this new standard also prompted Camst to improve SR communication on its website in order to increase the visibility of social initiatives. The company's online disclosure of SR was poor and still lacking with respect to all of the activities and strategies pursued. Also with reference to this aspect, Camst looks similar to other Italian companies. The survey conducted by Uni and Sodalitas (2014) indicates that companies do not give great visibility to ISO 26000's implementation. The communication strategy represents a critical aspect in a relevant number of companies.

Camst does not seem to differ from other companies nor does it require specific adaptations for ISO 26000's guidelines implementation related to its legal form and institutional model. As confirmed by the benefits identified, a cooperative can successfully adopt the standard and important results can be achieved even in the case of partial implementation.

Acknowledgments We would like to thank Paola Bertocchi and Ivano Minarelli from Camst as well as Claudia Strasserra, Sustainability Sector manager from Bureau Veritas Italia, for their valuable collaboration, sensitivity and support in providing us with the material necessary to develop the case study analysis. Moreover, we would like to thank the important suggestions provided by the reviewers.

References

Abdelkafi, N., & Täusher, K. (2015). Business models for sustainability from a system dynamics perspective. *Organization Environment, 29*(1), 74–96. https://doi.org/10.1177/1086026615592930.

Bailey, C. A. (2007). *A guide to qualitative field research* (2nd ed.). Newbury Park: Pine Forge.

Balzarova, M. A., & Castka, P. (2012). Stakeholders' influence and contribution to social standards development: The case of multiple stakeholder approach to ISO 26000 development. *Journal of Business Ethics, 111*(2), 265–279.

Bocken, N. M. P., Short, S. W., Rana, P., & Evans, S. (2014). A literature and practice review to develop sustainable business model archetypes. *Journal of Cleaner Production, 65*, 42–56.

Boström, M., & Hallström, K. T. (2013). Global multi-stakeholder standard setters: How fragile are they? *Journal of Global Ethics, 9*(1), 93–110.

Burlea Schiopoiu, A. (2013). Global environmental management initiative. In S. O. Idowu, N. Capaldi, L. Zu, & A. das Gupta (Eds.), *Encyclopedia of corporate social responsibility* (pp. 1241–1248). Heidelberg: Springer.

CAMST. (2015). *Bilancio Sociale CAMST 2015*. Bologna: Camst La ristorazione italiana.

De Deus, R. M., Seles, B. M. R. P., & Vieira, K. R. O. (2014). Organizations and ISO 26000: Review of concepts, motivators, and barriers to implementation [As organizações e a ISO 26000: Revisão dos conceitos, dos motivadores e das barreiras de implementação]. *Gestao e Producao, 21*(4), 793–809.

Eisenhardt, K. M. (1989). Building theories from case study research. *Academy of Management Review, 14*(4), 532–550.

Eisenhardt, K. M., & Graebner, M. E. (2007). Theory building from cases: Opportunities and challenges. *Academy of Management Journal, 50*(1), 25–32.

European Commission (EC). (2011, October 25). *A renewed EU strategy 2017–14 for corporate social responsibility*. COM(2011) 681 final, Brussels.

Gray, R., Adams, C., & Owen, D. (2014). *Accountability, social responsibility and sustainability*. Person: Accounting for society and the environment.

Habidin, N. F., Fuzi, N. M., Zamri, F. I. M., Hibadullah, S. N., & Desa, A. F. N. C. (2014). ISO 26000 efforts and corporate social responsibility performance in Malaysian automotive industry. *International Journal of Business Excellence, 7*(4), 515–529.

Hahn, R. (2013). ISO 26000 and the standardization of strategic management processes for sustainability and corporate social responsibility. *Business Strategy and Environment, 22*, 442–455.

Hahn, R., & Weidtmann, C. (2016). Transnational governance, deliberative democracy, and the legitimacy of ISO 26000: Analyzing the case of a global multistakeholder process. *Business and Society, 55*(1), 90–129.

Hemphill, T. (2013). The ISO 26000 guidance on social responsibility international standard: What are the business governance implications? *Corporate Governance: The International Journal of Business in Society, 13*(3), 305–317.

ISO. (2010, November 1). *Guidance on social responsibility, ISO 26000: 2010* (1st ed.). Final draft, Geneva.

ISO. (2012). *ISO 26000 post-publication survey 2012*. Geneva: ISO.

Johnston, A. (2011, August 29). *Constructing sustainability through CSR: A critical appraisal of ISO 26000*. University of Oslo Faculty of Law, Research paper no. 2011-33.

Katamba, D., Nkiko, C. M., Kazooba, C. T., Kemeza, I., & Mpisi, B. S. (2014). Community involvement and development: An inter-marriage of ISO 26000 and millennium development goals. *International Journal of Social Economics, 41*(9), 837–861.

Matacena, A. (2010). Corporate social responsibility and accountability: Some glosses. In M. G. Baldarelli (Ed.), *Civil economy, democracy, transparency and social and environmental accounting research role* (pp. 7–59). Milano: McGraw-Hill.

Moratis, L. (2016a). Out of the ordinary? Appraising ISO 26000's CSR definition. *International Journal of Law and Management, 58*(1), 26–47.

Moratis, L. (2016b). Signaling strategies for ISO 26000: A firm-level approach. *International Journal of Operations and Production Management, 36*(5), 512–531.

Moratis, L., & Cochius, T. (2011). *ISO 26000: The business guide to the new standard on social responsibility*. Sheffield: Greenleaf.

Mueckenberger, U., & Jastram, S. (2010). Transnational norm-building networks and the legitimacy of corporate social responsibility standards. *Journal of Business Ethics, 97*(2), 223–239.

Naumes, W., & Naumes, M. J. (2006). *The art and craft of case writing* (2nd ed.). London: M.E. Sharpe.

O'Riordan, L., & Fairbrass, J. (2008). Corporate social responsibility (CSR) models and theories in stakeholder dialogue. *Journals of Business Ethics, 83*(4), 745–758.

Patton, M. Q. (2002). *Qualitative research & evaluation methods* (3rd ed.). Thousand Oaks: Sage.

Perera, O. (2008). *How material is ISO 26000 social responsibility to small and medium-sized enterprises?* Winnipeg, Manitoba: International Institute for Sustainable Development.

Peršič, A., Markič, M., & Peršič, M. (2016). The impact of socially responsible management standards on the business success of an organisation. *Total Quality Management and Business Excellence, 29*(1–2), 225–237. https://doi.org/10.1080/14783363.2016.1174059

Ranängen, H., Zobel, T., & Bergström, A. (2014). The merits of ISO 26000 for CSR development in the mining industry: A case study in the Zambian Copperbelt. *Social Responsibility Journal, 10*(3), 500–515.
Schaltegger, S., Lüdeke-Freund, F., & Hansen, E. G. (2012). Business cases for sustainability: The role of business model innovation for corporate sustainability. *International Journal of Innovation & Sustainable Development, 6*(2), 95–119.
Schwartz, B., & Tilling, K. (2009). 'ISO-lating' corporate social responsibility in the organizational context: A dissenting interpretation of ISO 26000. *Corporate Social Responsability and Environemtnal Management, 16*, 289–299.
Smith, M. (2011). *Research methods in accounting* (2nd ed.). Thousand Oaks: Sage.
Toppinen, A., Virtanen, A., Mayer, A., & Tuppura, A. (2015a). Standardizing social responsibility via ISO 26000: Empirical insights from the forest industry. *Sustainable Development, 23*, 153–166.
Toppinen, A., Hänninen, V., & Lähtinen, K. (2015b). ISO 26000 in the assessment of CSR communication quality: CEO letters and social media in the global pulp and paper industry. *Social Responsibility Journal, 11*(4), 702–715.
UNI & Sodalitas. (2014). *UNI ISO 26000: la responsabilità sociale in concreto*. Milano: UNI.
UNI/PdR 18:2016. (2016, April 29). *Social responsibility in organizations - Guidance to the application of UNI ISO 26000*. ICS 03.100.01.
Valmohammadi, C. (2014). Impact of corporate social responsibility practices on organizational performance: An ISO 26000 perspective. *Social Responsibility Journal, 10*(3), 455–479.
von Weltzien Hoivik, H. (2011). Embedding CSR as a learning and knowledge creating process: The case for SMEs in Norway. *Journal of Management Development, 30*(10), 1067–1084.
Walker, T., & Beranek, F. (2015). Corporate resilience - 5th generation? In S. O. Idowu, N. Capaldi, M. Fifka, L. Zu, & R. Schimdpeter (Eds.), *Dictionary of corporate social responsibility* (pp. 118–119). Heidelberg: Springer.
Walker, T., & Schmidpeter, R. (2015). Maturity model of CSR. In S. O. Idowu, N. Capaldi, M. Fifka, L. Zu, & R. Schimdpeter (Eds.), *Dictionary of corporate social responsibility* (pp. 312–313). Heidelberg: Springer.
Yin, R. (2009). *Case study research: Design and methods* (4th ed.). Thousand Oaks, CA: Sage.
Zinenko, A., Rovira, M. R., & Montiel, I. (2015). The fit of the social responsibility standard ISO 26000 within other CSR instruments: Redundant or complementary? *Sustainability Accounting, Management and Policy Journal, 6*(4), 498–526.
Zompras, A., & Siakas, K. (2015). An investigation of ISO 26000 and social responsibility practices applied in IT companies. *International Journal of Human Capital and Information Technology Professionals, 6*(2), 33–48.

Mara Del Baldo is an Associate Professor of Small Business Management; Financial Accounting and Economics of Sustainability and Accountability at the University of Urbino Carlo Bo, Department of Economics, Society and Politics, School of Economics, Urbino, Italy.
Her main research interests include entrepreneurship and small businesses management; Corporate Social Responsibility; sustainability and business ethics; SMEs and networking strategies; accounting and gender; financial and integrated reporting; and social and environmental accounting research (SEAR). She is a member of: the European Council for Small Business, the Centre for Social and Environmental Accounting Research (CSEAR), the Italian Network on Business Reporting (NIBR; WICI), the SPES Forum, the Global Corporate Governance Institute (GCI), the Corporate Governance Experts Global Repository—Virtusinterpress and the European Business Ethics Network (EBEN) Italian Chapter. She is editorial board member and reviewer of several Italian and international scientific journals. She is also author of several scientific publications, including articles in Italian and foreign journals, chapters books, conference proceedings and books. She has given several lectures on invitation in various Italian and foreign universities (University of

Vigo, Spain; the Jurai Dobrila University of Pula; Croazia, the New Bulgarian University of Sofia, Bulgari; a the Corvinus University, Budapest, Hungaria and the University of Craiova, Romania).

Selena Aureli is an Assistant Professor of Financial Reporting at the University of Bologna at the Rimini Campus, Italy. Her latest articles cover diverse topics, such as inter-firm alliances, cooperation in tourism and also on network brand management, which have appeared in the International Journal of Economic Behavior, the International Journal of Management Cases and the European Journal of Tourism Research. Her research interests span widely, from entrepreneurship and small businesses, performance measurement and control systems, and mergers and acquisitions to financial and sustainability reporting. She is in the editorial board of three peer-reviewed journals: Piccola impresa-Small Business, Journal of Accounting and Management Information Systems and Management Control and serves as reviewer for several international academic journals. She collaborates with colleagues at the University of Urbino where she received her PhD in Business Administration and where she was previously enrolled as Assistant Professor of Accounting and Management Control.

ISO 26000: The Toyota Motor Corporation's CSR Approaches in India

Lars E. Isaksson and Nayan Mitra

1 Introduction

The relationship between businesses, organizations and societies are intertwined as they do not operate in a vacuum. Their operations in society and global environment is a critical factor in their ability to operate efficiently and effectively. This co-dependence has therefore evolved to become a part of overall corporate performance measurement. Corporate Social Responsibility (CSR) has therefore gained prominence in the regulatory sector where firms increasingly are expected to apply corporate transparency and report their CSR efforts (KPMG 2011; Isaksson 2012). If companies want to manage CSR and sustainability issues and obtain the trust of their stakeholders, they must communicate their efforts and provide concrete evidence of their commitments towards continual and long-term social improvement. It is for these reasons, it is necessary that any sustainability and responsibility oriented company define appropriate systems to implement, control, measure and evaluate their CSR performance. Over the years, many social and environmental standards and management solutions have been developed to evaluate and report the economic, social, environmental and sustainability performance of companies (Tencati 2013). The CSR movement is now entering a mainstreaming phase aided by standardization activities such as the Global Reporting Initiative (GRI); Environmental, Social and Governance (ESG); the AA1000 series and the ISO 26000 guide (Das Gupta 2012). While the GRI Sustainability Reporting Guidelines is a comprehensive, global, sustainability reporting standard (GRI 2014) and the AA1000 series (by the

L. E. Isaksson
Bond University, Gold Coast, Australia
e-mail: lars.isaksson@qut.edu.au

N. Mitra (✉)
Developmental Consultant, Kolkata, West Bengal, India

S. O. Idowu et al. (eds.), *ISO 26000 - A Standardized View on Corporate Social Responsibility*, CSR, Sustainability, Ethics & Governance,
https://doi.org/10.1007/978-3-319-92651-3_9

AccountAbility organization) is a principles-based operational guide aiming to help organizations become more sustainable and accountable, ISO 26000 helps clarify what CSR is, and what it can be for the individual firm, thus helping businesses and organizations translate principles into effective actions and share best practices relating to social responsibility, globally.

One company that started with their own policies and programs and now combine and enhance them with ISO 26000 is the Toyota Motor Corporation (TMC). The intention of this paper is to assess the 'Toyota Way' and its relation to, and compliance to the seven guiding principles of the ISO 26000 framework (Organizational Governance, Human Rights, Labour Practices, The Environment, Fair Operating Practices, Consumer Issues, and Community Involvement and Development).

2 The ISO 26000 Standard

The International Standardization Organization (ISO) has a long, advanced and reputable standing among national and multinational companies (MNCs). The ISO 26000:2010 (ISO 26000) standard is a framework to provide guidance on how businesses and organizations can operate in a socially responsible way. This means acting in an ethical and transparent way that contributes to the health and welfare of society (ISO 2017). While almost all other ISO standards (for instance the ISO 9001 Quality, the ISO 14001 Environmental Standard or the ISO 27001 Information Security standard) is certifiable, ISO 26000 instead provides guidance rather than requirements due to the non-universal approach to CSR. Social responsibility is simply too diverse and location dependent to be the same or even similar for any two companies (Isaksson and Woodside 2016). Hence, it helps clarify what CSR is, or can be, to support businesses and organizations translating certain principles into effective actions. It also shares the accumulated and sampled CSR best practices globally, representing an international consensus of multiple representatives from government, NGOs, industry, consumer groups and labour organizations. It therefore targets all types of organizations regardless of their industry belonging, size, for profit or for non-profit or geographical location (ISO 2017).

While the aforementioned ISO 14001 standard in itself is not a responsibility standard per se, it is a program that helps companies to better understand internal waste processes and hence perceived as one precursor to CSR. It is therefore widespread among firms taking sustainability seriously that Multinational Companies (MNCs) with an ISO 14001 certification also invest in general CSR programs or activities given the close relationship between the CSR and sustainability concepts. In fact, the mere presence of ISO 14001 enhance general CSR so that a firm's market value and/ or, attraction of perceived products or services is enhanced, which in turn can result in improved financial performance. It is however important to point out that while the more widespread ISO standard, ISO 9001 (a quality monitoring standard) is not in any way, related to CSR in general or ISO 26000 particularly, it is in some markets (predominantly in emerging economies) incorrectly communicated as a CSR component or substitute (Isaksson 2012). The reason for this misconception is that, quality, in itself, is a business-value oriented mechanism only, whereas standards for waste management or CSR can be both a driver of 'responsible performance' and overall business value. This is also documented in quantitative research findings. Isaksson (2012) found that MNCs' with advanced CSR standing applies industrial standards (e.g. the industrial standard ISO 14001 waste management or the ISO 9001 quality standard) to a greater extent when compared with other MNCs; and that one third of MNCs with existing ISO 14001 certification also intended to apply the ISO 26000 CSR-standard. It is therefore common in developed economies that MNCs holding an ISO 14001 certification also considers to implement ISO 26000 (Fig. 1; Table 1).

3 CSR Related Performance Opportunities

MNCs today engage in CSR to benefit from it. While this might indicate a level of cynicism, it is in effect a durable way to benefit all stakeholders including the company providing it; the reason being that CSR is considered to be a long-term investment that can lead to competitive and financial advantages (Carroll and Shabana 2010; Kang 2009; KPMG 2011; Orlitzky et al. 2003). Hence, if a company does something good for the society it will also be good for them. CSR is further claimed to have both a direct and indirect impact on firm performance (Wang et al. 2015) where the MNCs' level of financial performance can be improved by their CSR efforts with regards to primary measures like operating profit, sales growth, market share and return on assets (Isaksson 2012). Apart from the empirical support that CSR indeed provides positive financial performance (Wang et al. 2015; Orlitzky et al. 2003; Wood 2010), CSR can also have a positive influence on non-financial performance, for example via reduced idiosyncratic (firm specific) risk (Luo and Bhattacharya 2009). Even though it is difficult to measure, firms that deploy CSR can achieve market based intangible assets such as improved brand image and market reputation (Drucker 1984; Harrison et al. 2010; Moon and deLeon 2007; Waddock and Graves 1997; Wood 2010), positively affect employee courage, increase willingness towards customer service, which in turn can have a positive impact on customer loyalty and stability of cash flows (Fombrun 2000; Lev et al. 2011; Luo and Bhattacharya 2009; McWilliams and Siegel 2011). As a result, CSR is claimed to assist long-term stock performance and can be in the best interest of the shareholders (Luo and Bhattacharya 2009). In contrast, firms with poor CSR reputation, or poor CSR integration, can decrease their market value and even be hit harder in economic downturns (greater stock market decline) than firms with positive CSR reputation (Peloza 2006). While the above arguments provide general support for the business case of CSR, the new ISO 26000 CSR standard framework provides a roadmap to its achievement. This is especially important due to the implementation problems experienced by any company embracing CSR (Isaksson et al. 2014). ISO 26000 will for this reason aid many MNCs to mine the CSR related

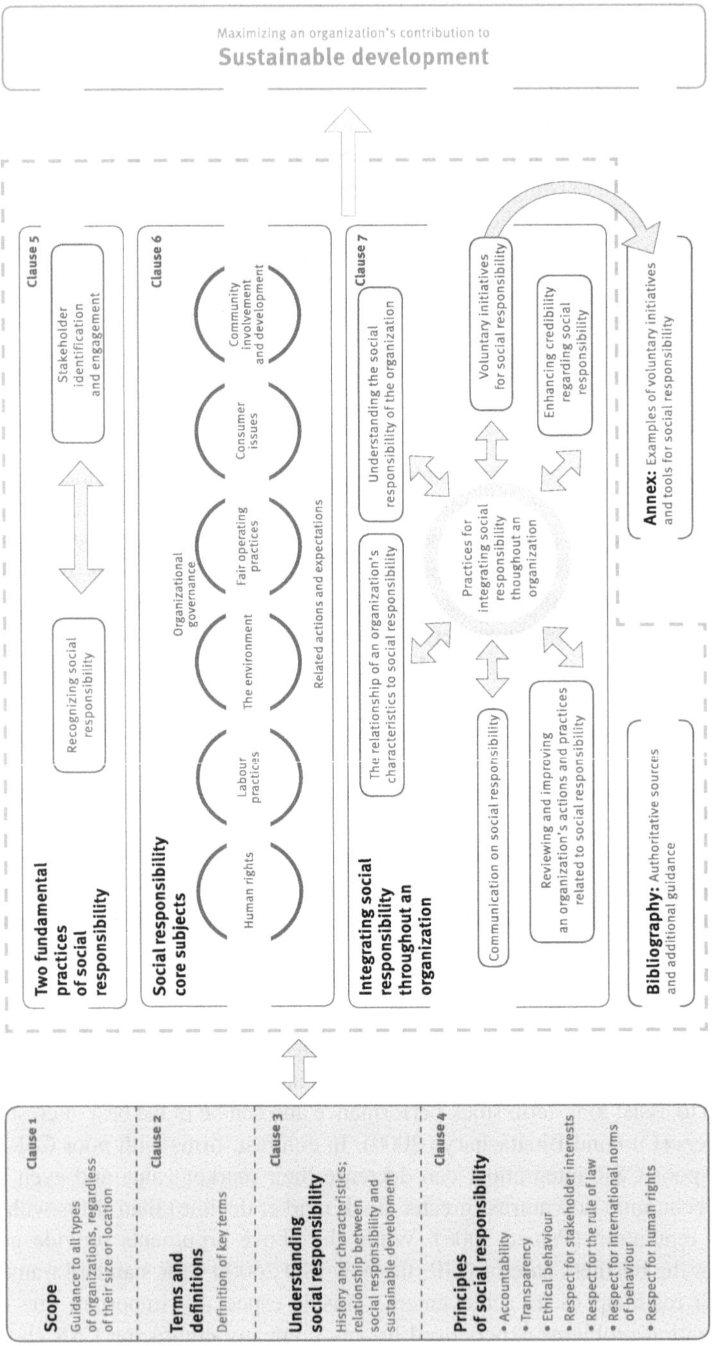

Fig. 1 Schematic Overview of ISO 26000. Source: ISO (2017)

Table 1 ISO 26000

Core subjects and issues	Clause
Core subject: Organizational governance	6.2
Core subject: Human rights	6.3
Issue 1: Due diligence	6.3.3
Issue 2: Human rights risk situations	6.3.4
Issue 3: Avoidance of complicity	6.3.5
Issue 4: Resolving grievances	6.3.6
Issue 5: Discrimination and vulnerable groups	6.3.7
Issue 6: Civil and political rights	6.3.8
Issue 7: Economic, social and cultural rights	6.3.9
Issue 8: Fundamental principles and rights at work	6.3.10
Core subject: Labour practices	6.4
Issue 1: Employment and employment relationships	6.4.3
Issue 2: Conditions of work and social protection	6.4.4
Issue 3: Social dialogue	6.4.5
Issue 4: Health and safety at work	6.4.6
Issue 5: Human development and training in the workplace	6.4.7
Core subject: The environment	6.5
Issue 1: Prevention of pollution	6.5.3
Issue 2: Sustainable resource use	6.5.4
Issue 3: Climate change mitigation and adaptation	6.5.5
Issue 4: Protection of the environment, biodiversity and restoration of natural habitats	6.5.6
Core subject: Fair operating practices	6.6
Issue 1: Anti-corruption	6.6.3
Issue 2: Responsible political involvement	6.6.4
Issue 3: Fair competition	6.6.5
Issue 4: Promoting social responsibility in the value chain	6.6.6
Issue 5: Respect for property rights	6.6.7
Core subject: Consumer issues	6.7
Issue 1: Fair marketing, factual and unbiased information and fair contractual practices	6.7.3
Issue 2: Protecting consumers' health and safety	6.7.4
Issue 3: Sustainable consumption	6.7.5
Issue 4: Consumer service, support, and complaint and dispute resolution	6.7.6
Issue 5: Consumer data protection and privacy	6.7.7
Issue 6: Access to essential services	6.7.8
Issue 7: Education and awareness	6.7.9
Core subject: Community involvement and development	6.8
Issue 1: Community involvement	6.8.3
Issue 2: Education and culture	6.8.4
Issue 3: Employment creation and skills development	6.8.5
Issue 4: Technology development and access	6.8.6
Issue 5: Wealth and income creation	6.8.7
Issue 6: Health	6.8.8
Issue 7: Social investment	6.8.9

Source: ISO (2017)

performance benefits and other operative improvements, the general objective being to increase the returns from their social responsible programs—doing more 'good' is reciprocal and creates a 'win-win' situation (Isaksson et al. 2014).

Two examples of MNCs in the process of implementing ISO 26000 is the global tooling company SANDVIK and the multinational insurance company SKANDIA. Like TOYOTA, SANDVIK embraces community involvement as part of their long-term business development since its inception 150 years ago. An example is their provision of education, health-supportive advice and medication regarding HIV and AIDS to their workers in Zambia, Africa, where this non-discriminating disease reached almost pandemic rates across the nation. SKANDIA shares this reciprocal view as described in their vision 'doing what is societal good is reciprocal and good for the company'. They therefore provide social solutions that benefits both the company (combining strategic and operative and financial objectives) and their target-market societies at large. For instance, by engaging in lobbying for legislative changes in building codes to decrease the risks of building fires and improved traffic environment (e.g. mandatory seat-belts for children). Improving the safety for the overall population also drives insurance sales and decrease insurance claims. To investigate how the ISO 26000 framework is implemented and operationalized, this study assess how the Toyota Motor Corporation (TMC) applies their CSR activities in that realm, by describing their global and local initiatives (in India).

4 Methodology

This Case study captures the research target objective (exemplifying how ISO 26000 can be applied) by analysing primary sources of information to provide genuine and valuable insight of an early implementation of the ISO 26000 'guided' standard. From a methodological perspective, this Case Study therefore goes beyond traditional 'secondary sources research' (for example, annual reports or reporting journalism alone) as it would not be holistic and limit the contribution to researchers and practitioners alike. The authors therefore met with both TMC and Toyota-Kirloskar (TK: the Indian Joint-Venture) representatives in person (in India) to conduct in-depth interviews, exchange research information and outcomes and engaged in a joint workshop as to validate current standing of the TMC policies, their local adaptations at TK and update management knowledge of what new insight, specialized CSR researchers potentially could contribute with.

An in-depth knowledge exchange meeting was prepared via pre-meeting correspondence during the first part of 2015 and finalized personally at Toyota's main factory in Bengaluru on the 26th of June 2015. This study is therefore rigid and robust in that the reported findings and statements were triangulated (i.e. using web-site material, 3rd party index and performance rankings, written correspondence exchange, personal site-visits and in-depth qualitative interviews) to validate this research. Apart from verifying previous findings, the interviews also yielded insight into strategic CSR decision making, the Top Management Team's (TMT's) role and managerial perceptions of reputation, intangible asset creation, and indirect financial performance. Other topics on the agenda were assessments about practical implementation and integration efforts and how CSR was communicated.

5 Toyota Motor Corporation's ISO 26000 Global Initiative

Considering that the ISO 26000 CSR standard, in contrast to the related standard ISO 14001 (environmental standard) is not certifiable, it is of interest to assess and exemplify how large MNCs implement their 'standardized' CSR efforts. One of the many MNCs that sincerely engage in CSR programs of the highest dignity is the Toyota Motor Corporation (TMC). Founded in 1937 in Japan by Mr. Sakichi Toyoda as a family owned business, TMC expanded its business over the coming decades to comprise 53 overseas manufacturing companies in 28 countries and regions. By the end of 2015, TMC was selling more than 30 different vehicle models across 170 countries. It is the legend of the Toyoda family's principles that siphons through the TMC, where they continue to strive, at all times, to be perceived as a trusted, respected and valued company to as many stakeholders as possible to share and buy into their mantras (Table 2). TMC's approach to achieve their high level of CSR is labelled the Toyota Way.

The Toyota Way has its roots in the 14 Principles (Table 3) which can be summarized by the two pillars supporting it: 'continuous improvement' and 'respect for people' (Liker 2004). We note that the Toyota Way applies the general sustainability concept as its anchor-point and concludes with the need to be a 'learning organisation' practicing 'continuous improvement' to achieve their long-term oriented philosophy (principle 14). It is the purpose of this paper to assess the Toyota Way and its relation to, and compliance to the seven guiding principles of the ISO 26000 framework (Organizational Governance, Human Rights, Labour Practices, The Environment, Fair Operating Practices, Consumer Issues, and Community Involvement and Development) in one of their foreign operations.

The current President of TMC (and Member of the Board of Directors), Mr. Akio Toyoda eloquently explains TMC's corporate view, philosophy and mindset regarding CSR as honouring the three mantras central to TMC: first, their motto "Yes, let's do it!" erected by the founder; second, developing their people as the foundation for business stability; and third, a focus on the Japanese '*genchi genbutsu*' code which highlights the necessity of on-site and hands-on experience (learning by doing). These foundational values make it necessary to focus on the market at hand, the current and future products, and how to interact with society in the markets they operate (Delmas and Toffel 2008; Gadenne et al. 2009). Their corporate vision is operationalized within its corporate culture and 'Charter of Corporate Behaviour' erected by the Japan Business Foundation, Nippon Keidanren, an alliance of leading Japanese corporations.

TMC therefore makes serious endeavours to resolve a range of social issues as part of the Toyota Global Vision. Let us take the example of their vehicles segment. On

Table 2 Five main principles of Toyoda

1. Always be faithful to your duties, thereby contributing to the company and the overall good
2. Always be studious and creative, striving to stay ahead of the times
3. Always be practical and avoid frivolousness
4. Always strive to build a homelike atmosphere at work that is warm and friendly
5. Always have respect for spiritual matters, and remember to be grateful at all times

Table 3 The 14 Toyota way principles

Section I: Long-term philosophy
Principle 1: Base your management decisions on a long-term philosophy, even at the expense of short-term financial goals
Section II: The right process will produce the right results
Principle 2: Create Continuous process flow to bring problems to the surface
Principle 3: Use "pull" systems to avoid overproduction
Principle 4: Level out the workload (*Heijunka*), i.e. 'Work like the tortoise, not the hare'
Principle 5: Build a culture of stopping to fix problems, to get quality right the first time
Principle 6: Standardized tasks are the foundation for continuous improvement and employee empowerment
Principle 7: Use visual control so no problems are hidden
Principle 8: Use only reliable, thoroughly tested technology that serves your people and processes
Section III: Add value to the organization by developing your people and partners
Principle 9: Grow leaders who thoroughly understand the work, live the philosophy, and teach it to others
Principle 10: Develop exceptional people and teams who follow your company's philosophy
Principle 11: Report your extended network of partners and suppliers by challenging them and helping them improve
Section IV: Continuously solving root problems drives Organizational Learning
Principle 12: Go and see for yourself to thoroughly understand the situation (*Genchi genbutsu*)
Principle 13: Make decisions slowly by consensus, thoroughly considering all options; implement decisions rapidly
Principle 14: Become a learning organization through relentless reflection (*Hansei*) and continuous improvement (*Kaizen*)

Source: Liker (2004)

the product-market level, TMC develops socially sustainable products beyond the current hybrid platform used in the Prius cars. In 2014, TMC launched the MIRAI fuel cell cars after two decades of research and development in response to global warming. By focusing on 'always better cars,' TMC is committed to repeatedly engage in new, more renewable and more sustainable platforms, for instance, the new hydrogen-based technology released in 2015. On the social-interaction level, TMC participates in socially responsible conceptual developments under the motto 'enriching lives of communities'. TMC therefore embraces a responsibility for each of the communities they do business in. After all, cars, traffic safety, the environment and other social issues are intertwined with each other and are part of any society's overall infrastructure. By cooperating and creating long-term trust based relationships with stakeholders of all categories (customers, residents, non-profit organizations, universities and various experts), TMC's CSR philosophy manifests in the most wanted state of affairs—harmony. It is therefore not a coincidence that TMC's CSR philosophy is expressed as "*to seek harmony with people, society and the global environment, and sustainable development of society through manufacturing*" (75 years of Toyota). The last part of this philosophy 'manufacturing' is in Japanese referred to as '*monozukuri'*. This aspect echoes a belief that technology and products manufactured in scale and scope holds the solutions for many societal challenges including the issue of environmental pollution. Since its foundation, TMC also applies the concept of '*kaizen'* (continuous improvement) to contribute to the sustainable development of society through the manufacture and provision of leading and innovative high quality products. While cars of course expand the freedom of mobility, they are also related to and affect a number of social and environmental issues. Hence, their CSR philosophy conveys a pursuit of harmonious business and the sustainable development of society through *monozukuri*. As an industrial group, and in accordance to general Japanese conglomerate aspirations, TMC also have business interests in unrelated industries. For example, in biotechnology, afforestation and energy—these are all in line with *monozukuri*. The basis for this rationale also leans on the primary, or head policy, of "contribution towards sustainable development". While there are many guiding principles within TMC, the principle regarding 'Social Contribution,' highlight this approach. Or expressed differently, wherever they do business, TMC actively promotes and engages, individually and with partners, in social contribution activities that help strengthen communities and contribute to the general enrichment of society at large. This approach has proven to be successful for TMC being recognized for genuine CSR efforts, for instance via inclusion in the 'Dow Jones Sustainability Index 2015; the 'FTSE4GOOD Index 2015' and the 'Morningstar Socially Responsible Investment Index 2014,' all with regards to their systematic ISO 26000 approach (Table 4).

At the TMC Japanese headquarter, CSR in the form of ISO 26000, was, until 2014, coordinated and promoted via a CSR committee, where the committee was responsible for legal compliances and decisions regarding 'which' social contribution activities to conduct and 'what' environmental initiatives to pursue. In effect, it is an approach to prioritize CSR activities in the light of the 'why to engage' and the 'what to engage in'. Due to the increased value that MNCs, consumers, governments and other stakeholders attach to CSR, countless MNCs around the world have increased their CSR efforts but also their Return on Investment (ROI) demands from CSR derivable strategic, corporate and market value. In response to these demands, TMC made organizational changes accordingly (TMC 2015). Discussions previously held by the corporate CSR Committee (for instance, corporate sustainability policies, social contribution, global CSR policies, CSR activities, ethical behaviour and compliance) are now the responsibility of the Corporate Planning Meetings and Corporate Governance Meetings. With the oversight of the Board of Directors, the Corporate Planning Meetings discuss growth strategies that incorporate the values that TMC provides to the society with regards to a variety of social

Table 4 ISO 26000 comparison within Toyota Motor Corporation: core subjects and issues in ISO 26000 and link to related initiatives

Core Topics in ISO 26000	Issues	Toyota's efforts
Organizational governance	1. Organizational governance	Society • CSR policy/organization and structure • Toyota global vision • CSR management based on the Toyota visionary management indices • corporate governancecompliance • Risk management
Human rights	2. Due diligence 3. Human rights risk situations 4. Avoidance of complicity 5. Resolving grievances 6. Discrimination and vulnerable groups 7. Civil and political rights 8. Economic, social and cultural rights 9. Fundamental principles and rights at work	Society • Respect for human rights • Collaboration with business partners • Employees • Compliance
Labour practices	10. Employment and employment relationships 11. Conditions of work and social protection 12. Social dialogue 13. Work health and safety 14. Human development and training in the workplace	Society • Special feature 03 stable base of employees
The environment	15. Prevention of pollution 16. Sustainable resource use 17. Climate change mitigation and adaptation 18. Protection of the environment, biodiversity and restoration of natural habitats	Environment • Contribution to a low carbon society • Contribution to a recycling-based society • Environmental protection and contribution to a harmony with nature society
Fair operating principles	19. Anti-corruption 20. Responsible political involvement 21. Fair competition 22. Promoting social responsibility in the value chain 23. Respect for property rights	Society • Collaboration with business partners • Compliance
Consumer issues	24. Fair marketing, factual and unbiased information and fair contractual practices	Society • Special feature 01—always better cars

(continued)

Table 4 (continued)

Core Topics in ISO 26000	Issues	Toyota's efforts
	25. Protecting consumers' health and safety 26. Sustainable consumption 27. Consumer service, support, and complaint and dispute resolution 28. Consumer data protection and privacy 29. Access to essential services 30. Education and awareness	• Special feature 02—enriching lives of communities • Initiatives for improving traffic safety • Customer first and quality first measures • Collaboration with business partners • Compliance Environment • Contribution to a low carbon society • Contribution to a recycling-based society • Social contribution activities • Environment • Traffic safety • Education • Society and culture
Community involvement and development	31. Community involvement 32. Education and culture 33. Employment creation and skills development 34. Technology development and access 35. Wealth and income creation 36. Health 37. Social investment	Society • Special feature 02—enriching lives of communities • Initiatives for improving traffic safety • Creating the future society Social contribution activities • Environment • Traffic safety • Education • Society and culture • Volunteer activities

Source: TMC (2015)

issues and to promote corporate wide CSR. They further integrate these corporate values with other managerial aspects, for example, with regards to governance structures, strategy realization and operative supervision. This is also in line with what other MNCs located in developed economies are doing (Table 5).

6 Toyota in India

While TMC engages in generic CSR (for instance responsible products like the Prius car) they also engage in specific CSR activities in line with the ISO 26000 program in several locations. In India for example, TMC operates via a joint venture (JV) erected with the Kirloskar Group which operates under the name Toyota-Kirloskar (TK). This JV operates out of the Indian Toyota automotive factory located in Bengaluru in

Table 5 CSR policy: contribution towards sustainable development

Category	Statement	ISO 26000 Ref. #:
TMC commitment	We, the Toyota Motor Corporation and our subsidiaries, take initiative to contribute to harmonious and sustainable development of society and the earth through all business activities that we carry out in each country and region, based on our guiding principles. We comply with local, national and international laws and regulations as well as the spirit thereof and we conduct our business operations with honesty and integrity. In order to contribute to sustainable development, we believe that management interacting with its stakeholders as described below is important, and we will Endeavour to build and maintain sound relationships with our stakeholders through open and fair communication. We expect our business partners to support this initiative and act in accordance with it.	1 2 4 22 23 24
Customers	1. Based on our philosophy of "customer first," we develop and provide innovative, safe and outstanding high quality products and services that meet a wide variety of customers' demands to enrich the lives of people around the world (guiding Principles 3 and 4).	25, 27 29, 30
	2. We will Endeavour to protect the personal information of customers and everyone else we are engaged in business with, in accordance with the letter and spirit of each country's privacy laws (guiding principles 1).	24, 28
Employees	3. We respect our employees and believe that the success of our business is led by each individual's creativity and good teamwork. We stimulate personal growth for our employees (guiding principles 5).	14
	4. We support equal employment opportunities, diversity and inclusion for our employees and do not discriminate against them (guiding principles 5).	5, 6, 10
	5. We strive to provide fair working conditions and to maintain a safe and healthy working environment for all our employees (guiding principles 5).	11, 13
	6. We respect and honour the human rights of people involved in our business and, in particular, do not use or tolerate any form of forced or child labour (guiding principles 5).	3, 4, 9
	7. Through communication and dialogue with our employees, we build and share the value "mutual trust and mutual responsibility" and work together for the success of our employees and the company. We recognize our employees' right to freely associate, or not to associate, complying with the laws of the countries in which we operate (guiding principles 5).	5, 7 8, 12
	7. Management takes a leadership role in fostering a corporate culture, and implementing policies, that promote ethical behaviour (guiding principles 1 and 5).	19, 20

(continued)

Table 5 (continued)

Category	Statement	ISO 26000 Ref. #:
Business partners	8. We respect our business partners such as suppliers and dealers and work with them through long-term relationships to realize mutual growth based on mutual trust (guiding principles 7).	21
	9. Whenever we seek a new business partner, we are open to any and all candidates, regardless of nationality or size, and evaluate them based on their overall strengths (guiding principles 7).	37
	10. We maintain fair and free competition in accordance with the letter and spirit of each country's competition laws (guiding principles 1 and 7).	21
Shareholders	11. We strive to raise corporate value while achieving a stable and long-term growth for the benefit of our shareholders (guiding principles 6).	–
	12. We provide our shareholders and investors with timely and fair disclosure of our operating results and financial condition (guiding principles 1 and 6).	1
Global society and local communities	*Environment*	
	13. We aim for growth that is in harmony with the environment by seeking to minimize the environmental impact of our business operations, such as by working to reduce the effect of our vehicles and operations on climate change and biodiversity. We strive to develop, establish and promote technologies enabling the environment and economy to coexist harmoniously, and to build close and cooperative relationships with a wide spectrum of individuals and organizations involved in environmental preservation (guiding principles 3).	15, 16 17, 18
	Community	
	14. We implement our philosophy of "respect for people"by honouring the culture, customs, history and laws of each country (guiding principles 2).	2, 7, 8
	15. We constantly search for safer, cleaner and superior technologies that satisfy the evolving needs of society for sustainable mobility (guiding principles 3 and 4).	26, 34
	16. We do not tolerate bribery of or by any business partner, government agency or public authority and maintain honest and fair relationships with government agencies and public authorities (guiding principles 1).	19, 20
	Social contribution	
	17. Wherever we do business, we actively promote and engage, both individually and with partners, in social contribution activities that help strengthen communities and contribute to the enrichment of society (guiding principles 2).	31, 32 33, 35 36, 37

Source: TMC (2015)

the Karnataka state region. TK operates TMC's global CSR aspirations locally with respect to the ISO 26000 principles and requirements (organizational governance, human rights, labour practices, fair operating practices, the environment, consumer issues, and community involvement and development). Their joint efforts have proven to be successful, recognized and awarded. For example, in 2013 TK won the 'Asia's Best CSR Practices Award'; in 2015 they won the 'Golden Peacock Award' to name a few.

This tactic means that CSR directives generated at the TMC's corporate headquarter in Japan are applied via locally adopted practices that better 'fit' the needs in a local market, or Bengaluru in this case. TMC thereby responds to their global CSR commitment by adapting and adjusting it to local market conditions. Being locally represented (by its partner TK) also means a closer understanding of local stakeholder needs. In fact, their CSR efforts goes beyond the Indian CSR legislative initiative (Section 135 of the 2013 Companies Act) as TK currently spends approximately 5% on local CSR due to the profound generated benefits for their Indian operations. The Act stipulates that every company that reaches the inclusion criteria must spend at least 2% of their average net profit (average of the previous 3 years) on CSR activities in India (Table 6). Since even the most serious CSR activity, risk to be unsuccessful (for example, where there is a discrepancy, or a misalignment, between local need and provided activity); therefore the truer a company is to their CSR aspirations and beliefs, the more willing they are to design and implement CSR correctly and to adopt it to the local conditions.

By reviewing how the ISO 26000 standard are perceived and applied within TMC through the examples of locally applied CSR initiatives in Bengaluru (by TK), we provide valuable insight into the operationalization of the ISO 26000 framework. When for instance TK in June 2016, announced a proactive high-impact campaign to ensure 100% sanitation (for example widespread access to toilets, water and sewage) in schools and villages in the Karnataka region in India (the region where TK's factory workers reside), they actively announced a battle to fight one of the grave concerns in India—open defecation. Being a serious short and long term threat to human health, open defecation has a severe effect on society, employees and company, where everybody is on the 'losing' side. If people are sick and continue to spread defecation related diseases, the result is a far reaching domino-like effect of

Table 6 Key Highlights of the Section 135 of the Companies Act, 2013 (India)

• Every company covered by the Act's inclusion criteria must spend at least 2% of their average net profit in the previous 3 years on CSR activities.
• The law applicable from the fiscal year 2014–2015 require companies to erect a CSR committee consisting of their board members, including one or more independent directors.
• The Act applies to companies with: • an annual turnover exceeding 1000 crore INR (approx. US$160 million), or • a net worth exceeding 500 crore INR (approx. US$80 million), or • a net profit exceeding 5 crore INR (approx. US$0.8 million); where 1 US$ =68.01 INR as on April 1, 2018.

sick children leading to sick workers leading to lower productivity at the factory level. This is of paramount importance considering that a 10% productivity cut represents an annual drop in sales of approximately 14,200 cars (Economic Times India 2017). In contrast, if children and factory workers are overall healthy and not sick, they are able to work in the factory and their children attend schooling i.e. provide for their families and society, and subsequently create a win-win-win scenario for the local society, the local contractors and builders, the company and the market place. TK's comprehensive project has been rolled out across 102 schools in 88 villages in Magadi, Kanakapura and Ramanagara, with the target of eliminating open (public) defecation in at least 40 villages across Karnataka.

From an ISO 26000 perspective, this program is related to both the 'Labour Practices' category, Issue 11, 12 and 13 (Conditions of work and social protection, Social dialogue and Health and safety at work respectively) and the 'Community Involvement and Development' category, Issue 36 and 37 (Health and Social investment).Another local example (June 2016) is to support, enable and motivate school children to remain within the bosom of the educational realm. To accomplish this, TK distributed books and school bags to 11,000 school children across 180 schools in Bidadi. This occasion marked the 13th year of Toyota's ongoing efforts to support the local community at the grass root level. The Vice President of External Affairs, CSR & Public Relations of TK, personally handed over the books and bags to local school children to lead by example, thereby displaying to other companies that at least one large MNC sincerely embraces CSR. From an ISO 26000 perspective, this program is related to the 'Human Rights' category 3 and 8 (Human rights risk situations and Economic, social and cultural rights) and the 'Community Involvement and Development' category, Issue 31 and 32 (Community involvement and Education and culture).

The 'Fair Operating Practices' is another core topic of TMC's ISO 26000 approach. This practice targets rampant local corruption and is a robust example of locally adopted corporate policy. The 'Anti-corruption' and 'Fair competition' categories (issue 19 and 21 respectively), is applied within TK in form of a whistle-blowers policy (Fig. 2). This locally created, erected and enforced policy is viewed as a necessity to overcome everyday threats to long-term corporate objectives as well as to fight petty crime nuisance among all managerial levels and employees—all to support general stakeholders via employee involvement and empowerment. This whistle-blower policy entails financially related fraud, bribery and a range of ethically related issues. This is of particular importance in India suffering from endemic and complex bureaucracy and omnipresent corruptive behaviour. The former is evident in India's overall 'ease-of-doing-business' rank 130th out of 191 nations (DoingBusiness Organisation 2017) and the latter where India rank 79th out of 176 countries (Transparency 2017), in par with less developed economies like Belarus and Albania. For a company like TMC and its JV partner TK, erecting a factory building and solving legal disputes are nightmarish with a rank of 185th for dealing with construction permits and 172nd in the world for enforcing contracts (Table 7).

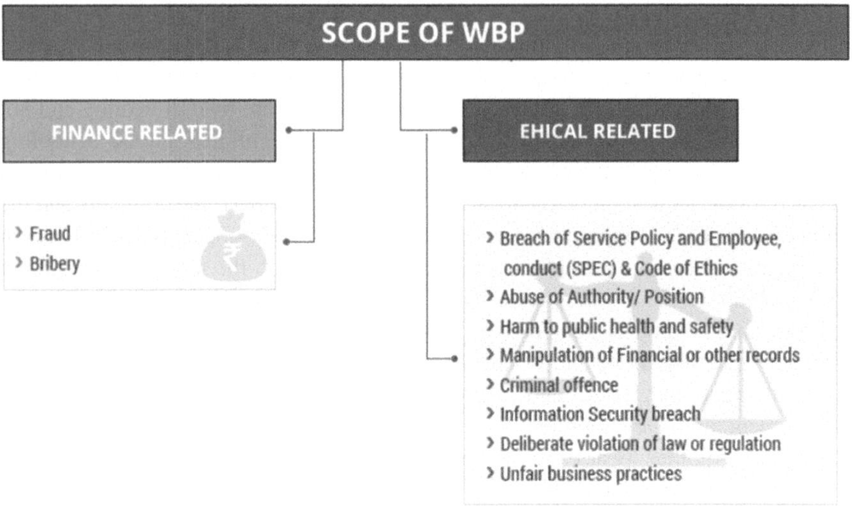

Fig. 2 TMC Whistle-Blower policy. Source: TMC (2015)

Table 7 Schedule VII of the companies act, 2013 (India)

(i) Eradicating hunger, poverty and malnutrition, promoting preventive health care and sanitation, including contribution to the Swachh Bharat Kosh set-up by the Central Government for the promotion of Sanitation and making available safe drinking water;
(ii) Promoting education, including special education and employment enhancing vocational skills especially among children, women, elderly, and the differently-abled and livelihood enhancement projects;
(iii) Promoting gender equality, empowering women, setting up homes and hostels for women and orphans; setting up old age homes, day care centres and such other facilities for senior citizens and measures for reducing inequalities faced by socially and economically backward groups;
(iv) Ensuring environmental sustainability, ecological balance, protection of flora and fauna, animal welfare, agro-forestry, conservation of natural resources and maintaining quality of soil, air and water, including contribution to the Clean Ganga Fund set-up by the Central Government for the rejuvenation of river Ganga;
(v) Protection of national heritage, art and culture including restoration of buildings and sites of historical importance and works of art; setting up public libraries; promotion and development of traditional arts and handicrafts;
(vi) Measures for the benefit of armed forces veterans, war widows and their dependents;
(vii) Training to promote rural sports, nationally recognised sports, para Olympic sports and Olympic sports;
(ix) Contributions or funds provided to technology incubators located within academic institutions which are approved by the Central Government;
(x) Rural development projects;
(xi) Slum development.

While TMC and TK jointly apply voluntary CSR in accordance to the TMC corporate philosophy and beliefs, their implementation of the industrial CSR standard ISO 26000 means that they, in the process, comply with Indian legislation as regulated by the 2013 Companies Act. Section 135 of the Companies Act 2013 is the formalised document that brought CSR from the 'backroom to the boardroom' (Mitra and Schmidpeter 2016) by placing the responsibility of CSR compliance and reporting to the Board or the Top Management Team (TMT) of mandated companies. One noteworthy implication is that TMC in their approach also follows recommendations that strategic customization of CSR initiatives maximizes CSR ROI (Isaksson et al. 2014). Moreover, their interventions in Sanitation and promotion of Education, also falls within the purview of the 'Schedule VII' of the Companies Act 2013 that helps prioritise and customize CSR applications for rapid results and increases the possibility of yielding high impact outcomes while maintaining, servicing and honouring stringent CSR norms and goals. It is expected that both TMC and TK will continue to contribute societal improvements beyond what they already accomplished, as more than 700,000 people have seen their lives improved across India, and the results are prominent. Toyota-Kirloskar ended the fiscal year 2016–2017 with annual sales revenue growth of 11%. Regardless to what extent their CSR efforts (i.e. the combination of legislative compliance and genuine beliefs that societal contribution is beneficial) contributes to the result, the new fiscal year continues to display signs of success. Sales revenue for the month of March 2017 for example, were reported to increase by 80% compared to the same month previous year (Economic Times 2017) (Table 8).

7 Conclusion

This case study assessed how the ISO 26000 Corporate Social Responsibility guiding framework can be implemented and operationalized. This is important for three distinct reasons. First, since businesses, organizations and societies are intertwined, their reciprocal relationship and services exchange needs to progress beyond tax payments (company to society) and provision of labour (society to company). Second, despite CSR being a hot topic, it is increasingly difficult to successfully apply due to the lack of universality. Third, since CSR has gained weight in the regulatory sector and firms increasingly are expected to be transparent and report their CSR activities, ISO 26000 is viewed as *the* tool for its achievement despite being a guiding framework only, in contrast to certifiable and universal standards.

Among the benefits of this investigation, we commence by highlighting that the robust and triangulated operative findings act as an example of 'how-to' generally apply CSR in a local context (here India) as well as a benchmark of 'how-to' comply with the Indian CSR legislation specifically. By discussing the award winning 'Toyota Way' and its relation to, and compliance to the seven guiding principles of the ISO 26000 framework (Organizational Governance, Human Rights, Labour

Table 8 Social Contributions by TMC and TK, 2015

Focus Area	Activity	Target Recipient	Coverage Area	Reach
Skill development	TTTI	School children	Bidadi	252
Skill development	TTEP	ITI	19 institutes across India	3518
Road safety	TSEP	School children	5 locations	650,000
Education	Books and bags	Govt. school children	8 gram panchayat of Bidadi Hobli	8500
Environment	Competition	School children	Bidadi Hobli	500
Environment	NDTV-Greenathon	Stakeholders	Nation wide	(villages) 1170
Health and hygiene	Water purification	Local community	Ittamadu and Byramangala	16,000
Health and hygiene	Sanitation	Individual households	Manchanyakanalli, Ittamadu, Byramangala	(families) 450
Health and hygiene	Health camps	General public and school children	Manchanyakanalli, Ittamadu, Byramangala, Bidadi	2140
Community development	Reconstruction of Ashrama Patashale (orphanage residential school)	School children	Byramangala	55
Community development	Water tank (large)	General public	Ramnagara	(population) 50,000
Community development	Water tank (small)	General public	Manchanyakanalli	(families) 750
Community development	Reconstruction of school	School children	Bidadi	50
			Total lives touched:	729,852

Source: TMC (2015)

Practices, The Environment, Fair Operating Practices, Consumer Issues, and Community Involvement and Development), this investigation further provides MNCs' with a systematic fashion to design and deliver genuine value for both the provider (here Toyota) and the receiver (the society). It is also important to clarify that while ISO 26000 framework provides a roadmap for systematic implementation of CSR, the most important contribution is its support that aid businesses to translate their CSR principles and policies into actionable win-win-win programs.

While these are substantial contributions, there are also important shortcomings suggested for future researchers to address. First, this paper only investigates one company—Toyota. It further only studies CSR efforts in one country (India) and hence lacks universality. Third, the investigation lack empirical strength being a case based study. Finally, it would be beneficial to verify measureable gains—'returns' for the provider to further strengthen the business case for legislated CSR.

References

75 years of Toyota. Retrieved June 20, 2016., from http://www.toyota-global.com/company/history_of_toyota/75years/data/conditions/philosophy/overview.html

Carroll, A. B., & Shabana, K. M. (2010). The business case for corporate social responsibility: A review of concepts, research and practice. *International Journal of Management Reviews, 12*(1), 85–105.

Das Gupta, A. (2012). Corporate social responsibility and strategy: A bird's eye view (article). *Global Business Review, 13*(1), 153–165.

Delmas, M. A., & Toffel, M. W. (2008). Organizational responses to environmental demands: Opening the black box. *Strategic Management Journal, 29*(10), 1027–1055.

DoingBusiness Organisation. (2017). *Economic ranks of India*. Retrieved April 20, 2017, from http://www.doingbusiness.org/data/exploreeconomies/india

Drucker, P. F. (1984). The new meaning of corporate social responsibility. *California Management Review, 26*, 53–63.

Economic Times India. (2017). *Toyota-Kirloskar sales FY 2016-2017*. Retrieved April 25, from http://economictimes.indiatimes.com/news/industry/auto/news/industry/toyota-sales-up-80-in-march-closes-fy17-with-11-growth/articleshow/57942402.cms

Fombrun, C. J. (2000). Opportunity platforms and safety nets: Corporate citizenship and reputational risk. *Business and Society Review, 105*(1), 85.

Gadenne, D., Kennedy, J., & McKeiver, C. (2009). An empirical study of environmental awareness and practices in SMEs. *Journal of Business Ethics, 84*(1), 45–63.

Harrison, J. S., Bosse, D. A., & Phillips, R. A. (2010). Managing for stakeholders, stakeholder utility functions, and competitive advantage. *Strategic Management Journal, 31*(1), 58–74.

Isaksson, L E. (2012, December 11). *CSR: A study of strategic management and firm performance*. Doctoral dissertation, Bond University, Australia.

Isaksson, L. E., & Woodside, A. (2016). Modelling firm heterogeneity in corporate social performance and financial performance. *Journal of Business Research, 69*, 3285–3314.

Isaksson, L., Kiessling, T., & Harvey, M. (2014). Corporate social responsibility: Why bother? *Organizational Dynamics, 43*(1), 64–72.

ISO. (2017). *ISO26000 standard framework for CSR*. Retrieved April 20, 2017, from https://www.iso.org/obp/ui/#iso:std:iso:26000:ed-1:v1:en

Kang, J. (2009). Corporate social responsibility? Not my business any more: The CEO horizon problem in corporate social performance. *Academy of Management Perspectives, 2009*(1), 1–6.

KPMG. (2011). KPMG international survey of corporate responsibility reporting 2011. *KPMG, 1*(1).

Lev, B., Petrovits, C., & Radhakrishnan, S. (2011). Is doing good for you? How corporate charitable contributions enhance revenue growth. *Strategic Management Journal, 31*(2), 182–200.

Liker, J. K. (2004). *The Toyota way: 14 management principles from the world's greatest manufacturer* (pp. 37–41). New York: Tata McGraw-Hill.

Luo, X., & Bhattacharya, C. B. (2009). The debate over doing good: Corporate social performance, strategic marketing levers, and firm-idiosyncratic risk. *Journal of Marketing, 73*, 198–213.

McWilliams, A., & Siegel, D. S. (2011). Creating and capturing value. *Journal of Management, 37*(5), 1480–1495.

Mitra, N., & Schmidpeter, R. (2016). The why, what and how of the CSR mandate: The India story. In *Corporate social responsibility in India: Cases and developments after the legal mandate*. Cham: Springer.

Moon, S.-G., & deLeon, P. (2007). Contexts and corporate voluntary environmental behaviors: Examining the EPA's green lights voluntary program. *Organization Environment, 20*(4), 480–496.

Orlitzky, M., Schmidt, F. L., & Rynes, S. L. (2003). Corporate social and financial performance: A meta-analysis. *Organization Studies, 24*(3), 403–441.

Peloza. (2006). Using corporate social responsibility as insurance for financial performance [article]. *California Management Review, 48*(2), 52–72.

Tencati, A. (2013). Corporate reporting frameworks. In G. Aras & D. Crowther (Eds.), *A handbook of corporate governance and social responsibility* (pp. 413–436). Surrey, GU: Gower Publishing.

TMC. (2015). *Interview at Bengaluru*. The TMC headquarter in India.

Transparency International. (2017). *Corruption perceptions index 2016*. Retrieved April 27, from http://www.transparency.org/news/feature/corruption_perceptions_index_2016

Waddock, S. A., & Graves, S. B. (1997). The corporate social performance-financial performance link. *Strategic Management Journal, 18*(4), 303–319.

Wang, Q., Dou, J., & Jia, S. (2015). A meta-analytic review of corporate social responsibility and corporate financial performance: The moderating effect of contextual factors. *Business and Society, 58*(8), 1083–1121. https://doi.org/10.1177/0007650315584317.

Wood, D. J. (2010). Measuring corporate social performance: A review. *International Journal of Management Reviews, 12*(1), 50–84.

Additional Resources

Toyota in India. *Corporate*. Retrieved May 22, 2016, from http://www.toyotabharat.com/toyota-in-india/about/

Toyota in India. *CSR*. Retrieved May 22, 2016, from http://www.toyotabharat.com/toyota-in-india/social-contribution/

Toyota in India. *Sustainability*. Retrieved May 22, 2016, from http://www.toyotabharat.com/environment/env-management/

Toyota Motor Corporation. Retrieved June 10, 2016, from http://www.toyota-global.com/

Toyota Motor Corporation. *CSR program*. Retrieved June 10, 2016, from http://www.toyota-global.com/sustainability/csr/index.html

Toyota Motor Corporation. *Sustainability*. Retrieved June 10, 2016, from http://www.toyota-global.com/sustainability/index.html

Lars E. Isaksson (Ph.D. Management- Global Strategy, MBA, MIB) is a lecturer at QUT, Australia. He has lectured subjects related to International Business, Strategy, Supply-Chain and Cross-Cultural management at the Undergraduate, Postgraduate, MBA and Executive MBA level. His research spans across global business strategy and Corporate Social Responsibility (CSR) in relation to strategic management subjects in multi-national companies and the economics of CSR. His research focuses on the effects of corporate behaviour and how organizations can improve their financial performance through reputation building and strategy-derived brand- and market improvements and performance.

Prior to his Ph.D., Dr. Isaksson worked for more than two decades in industry as a strategy consultant and solution sales specialist in various managerial and executive positions providing strategic improvements to companies such as Coca-Cola, ABB, Toyota, Carlsberg and IBM. His international industry exposure spans more than 60 countries in Europe, the Americas, the Middle East and Asia-Pacific.

Nayan Mitra (Ph.D. Management - Corporate Social Responsibility) comes with a rich mix of diverse professional experience of over 17 years. She straddles seamlessly between academics, social and corporate sectors. As a Developmental Consultant and Researcher, she works closely with some of the eminent Corporations and not-for profits of India as well as being in their Advisory and Board level capacity. She has conducted several social researches for multi-lateral agencies; the findings of which have become important bases for sustainable action. She has been a resource person in eminent Institutions of higher learning in the areas of CSR and Corporate Governance and has important peer reviewed research publications to her credit in double blind peer reviewed national and international Academic journals and books as well as delivered at key Conferences. Scientific committee member of a number of international consortium, her book, 'Corporate Social Responsibility in India: Cases and Development after the Legal Mandate' alongwith co-editor Dr. Rene Schmidpeter is a first book of its kind that charters the development of mandated Indian CSR from a multi-stakeholder perspective. She is also the recepient of the much coveted India CSR Author Award for two years in a row, 2017 and 2018; and spearheads the India CSR Leadership Series. She was a finalist of the prestigious Chevening Gurukul Scholarship for Leadership and Excellence—2013, as conferred by the Foreign and Commonwealth Office (FCO) of the British Deputy High Commission.

A Cost-Benefit Analysis of ISO's Standard on Social Responsibility: A Review and Appraisal

Mark Anthony Camilleri

1 Introduction

ISO 26000 provides guidance on social responsibility issues for businesses and other entities. This standard comprises issues relating to labor practices, conditions of employment, responsible supply chain management, responsible procurement of materials and resources, fair operating practices as well as recommendations for stakeholder engagement. ISO 26000 is aimed at all types of organizations, regardless of their activity, size or location. It clarifies what social responsibility is; it explains how businesses and other organizations can translate laudable principles into effective actions as they share best practices relating to social responsibility matters. ISO 26000's core subjects respect the international norms as the standard supports organizations to follow social responsibility principles on accountability, transparency, ethical behaviors and fair operating practices. Therefore, ISO 26000 supports different types of organizations in their engagement with stakeholders hailing from diverse contexts. ISO 26000 assists them in their socially responsible behaviors.

However, for the time being, there is still limited research and scant empirical evidence on ISO 26000 and on how the standard is actually being perceived by stakeholders (although there are a few exceptions, see Hahn 2013; Hahn and Weidtmann 2016; Marques 2012; Claasen and Roloff 2012; Castka and Balzarova 2008a, b). In this light, this chapter presents a thorough literature review on this promising subject. It shows that, very often ISO's social responsibility standard is guiding business in their stakeholder engagement and sustainability behaviors. This contribution suggests that companies ought to improve the relationships with suppliers and distributors in their value chain. In a nutshell, it demonstrates the costs and

M. A. Camilleri (✉)
University of Malta, Msida, Malta
e-mail: mark.a.camilleri@um.edu.mt

S. O. Idowu et al. (eds.), *ISO 26000 - A Standardized View on Corporate Social Responsibility*, CSR, Sustainability, Ethics & Governance,
https://doi.org/10.1007/978-3-319-92651-3_10

benefits for those organizations who are considering taking on board the ISO 26000 standard. In addition, a case study that has been drawn from ISO's own domain; shows how a British retail company, such as Marks and Spencer has stepped up in its commitment across its supply chain to create a fair workplace environment for other businesses and clusters. Therefore, this chapter contends that ISO 26000 is helping corporations to raise their profile among stakeholders as they follow the standards' guiding principles on social responsibility.

2 Background

Many social and environmental standards were developed before ISO 26000 have provided a platform for engagement with multiple stakeholders (Boström and Hallström 2010). Some examples of multi-stakeholder standards include the standards for sustainable forestry (FSC—Forestry Stewardship Council), sustainable fishing (MSC—Marine Stewardship Council) or the fair trade mark (by Fairtrade Labeling Organization International). These standards have typically emerged following numerous partnerships agreements between non-governmental organizations (NGOs) and large multinational corporations. The FSC standards were created through collaborative networks including Greenpeace, Rainforest Alliance, and Home Depot (Conroy 2007 in Balzarova and Castka 2012). Moreover, the Leadership in Energy and Environmental Design (LEED) is yet another example of a constructive partnership of the Natural Resources Defense Council and the Rocky Mountain Institute, among others.

In a similar vein, the ISO standards are documents that are established by consensus and approved by a recognized body, that provides, for common and repeated use, rules, guidelines or characteristics for activities or their results, aimed at the achievement of the optimum degree of order in a given context' (ISO 2004). Therefore, standards can be considered as recipes for what should be done by their adopters, while the standardization process can be defined as the production of specific rules (Brunsson and Jacobsson 2000). ISO 26000 was developed through constructive partnerships (Rangan et al. 2006) between private and public players as the International Standards Organization wanted to introduce its first un-certifiable standard. Several actors have been involved in developing this Social Responsibility standard. ISO 26000's working group contained 355 experts, 35 organizations, and 72 national representatives (Schwartz and Tilling 2009). Bowers (2006) observed that different stakeholders in ISO 26000 development have participated in the process of standard setting with clear positions in anticipation of its implementation. The industry stakeholders wanted to ensure that the social responsibility standard will be manageable for them. The International Labor Organization (ILO) was involved in the preliminary discussions as it safeguarded the employees' interests. Moreover, several governments were also involved in the preparations of ISO 26000. Other groups including consultants, academics, and certifiers hoped that a social responsibility standard would generate the same level of business that grew up around ISO 9001 and 14,001 (Balzarova and Castka 2012; Castka and Balzarova 2008b).

Multiple stakeholders have created links between the standard under development and other extant standards, guidelines or international instruments. The final version of the ISO 26000 standard contained a large list of voluntary initiatives and tools on social responsibility. These guidelines were proposed and pushed by the participating stakeholders. Therefore, the ISO 26000 standard-setting process involved stakeholder engagement among different groups.

Hence, the social responsibility standard involved a legitimization strategy that was crafted through a multi-stakeholder involvement with an emphasis on participatory decision making and democracy. This has inevitably led to common guidance on social responsibility concepts, definitions and methods of evaluation (Hahn and Weidtmann 2016; Fransen and Kolk 2007).

3 The Rationale Behind ISO 26000: The Standard on Social Responsibility

In 2010, the development of ISO 26000 has represented a milestone in multi-stakeholder standards development that supported the integration of social responsibility into management processes (Toppinen et al. 2015; Hahn 2013). Yet, ISO 26000 has never been considered as a management standard as its use cannot be certified unlike the earlier ISO standards, such as ISO 9000 and ISO 14001. The certification requirement has not been incorporated into the development and reinforcement process of ISO 26000 because industry representatives were concerned that costly certification requirements could overburden their businesses. Nevertheless, ISO's work item proposal for organizational social responsibility was intended to accomplish the following issues (Arzova 2009 in Idowu and Leal Filho 2009):

- Assist organizations in addressing their social responsibilities while respecting cultural, societal, environmental, and legal differences and economic development conditions;
- Provide practical guidance related to making social responsibility operational;
- Assist with identifying and engaging with stakeholders and enhancing credibility of reports and claims made about social responsibility;
- Emphasize performance results and improvement;
- Increase confidence and satisfaction in organizations among their customers and other stakeholders;
- Achieve consistency with existing documents, international treaties and conventions, and existing ISO standards;
- Promote common terminology in the social responsibility field;
- Broaden awareness of social responsibility;

- This standard is not intended to reduce government's authority to address the social responsibility of organizations.

ISO 26000 thus aims to help organizations manage their social responsibility. It helps to improve the individuals' working and living conditions, whilst fostering better opportunities to different organizations as they benchmark their social responsibility efforts. ISO 26000 has the potential to capture the context-specific nature of social responsibility. Even though the standard aims to unify and standardize social responsibility practices, it also acknowledges that organizations have a responsibility to bear as they are expected to address the strategic areas that are relevant to their business (Hahn 2013; Figge et al. 2002). Therefore, the ISO 26000 standard provides guidance on the integration of social responsibility into management processes and on matters relating to stakeholder engagement. Its core subjects and issues are presented hereunder in Table 1. ISO 26000 represents one of the most essential areas of social responsibility that an organization should take into consideration in order to maximize its contribution to sustainable development.

The standard's goal is to encourage organizations to adopt socially responsible approaches by reviewing their extant operating practices on organizational governance, human rights, labor practices, environment, fair operating practices, consumer issues and community involvement and development (ISO 2014). ISO 26000 provides guidance on stakeholder identification and engagement, it assists in improving social responsibility communications and it helps to integrate responsible business practices into strategies, systems and processes. Hence, ISO 26000 advises the practicing organizations to take into account their varied stakeholders' interests. The constructive partnerships agreements with multiple stakeholders are beneficial for the potential of effective consensus building, knowledge sharing, interest representation, and achievement of legitimacy (Fransen and Kolk 2007). According to Castka and Balzarova (2008a, p. 276), 'ISO 26000 aims to assist organizations and their network in addressing their social responsibilities—as they provide practical guidance that is related to operationalizing CSR, identifying and engaging with stakeholders and enhancing credibility of reports and claims made about CSR'. ISO 26000 can be viewed as an approach to CSR that is rooted in a quality management framework. Moratis (2015) has also reiterated the key contents and tenets of ISO 26000 as he examined strategies that could enhance the credibility of the corporations' social responsibility claims. He argued that the concept of credibility relates to skepticism, trust and greenwashing. Consequently, the organizations that are renowned for their CSR credentials will have a better reputation and image among stakeholders. This will result in significant improvements to the firms' bottom lines.

Berman et al. (1999) suggested that one approach to how organisations approach stakeholder management is based on an instrumental approach (strategic stakeholder management). They held that the organizations' concern toward stakeholders is motivated by their self-interest as they strive to improve their financial performance. Yet, there were several empirical studies that have often yielded contradictory results about whether social responsibility can bring financial returns (Camilleri 2012;

Table 1 Core subjects and issues in ISO 26000

Organizational governance	• Organizational governance
Human rights	• Due diligence
	• Human rights risk situations
	• Avoidance of complicity
	• Resolving grievances
	• Discrimination and vulnerable groups
	• Civil and political rights
	• Economic, social and cultural rights
	• Fundamental principles and rights at work
Labour practices	• Employment and employment relationships
	• Conditions of work and social protection
	• Social dialogue
	• Health and safety at work
	• Human development and training in the workplace
The environment	• Prevention of pollution
	• Sustainable resource use
	• Climate change mitigation and adaptation
	• Protection of the environment, biodiversity and
	• Restoration of natural habitats
Fair operating practices	• Anti-corruption
	• Responsible political involvement
	• Fair competition
	• Promoting social responsibility in the value chain
	• Respect for property rights
Consumer issues	• Fair marketing, factual and unbiased information and fair contractual practices
	• Protecting consumers' health and safety
	• Sustainable consumption
	• Consumer service, support, and complaint and
	• Dispute resolution
	• Consumer data protection and privacy
	• Access to essential services
	• Education and awareness
Community involvement and development	• Community involvement
	• Education and culture
	• Employment creation and skills development
	• Technology development and access
	• Wealth and income creation
	• Health
	• Social investment

ISO (2014)

Orlitzky et al. 2003; McWilliams and Siegel 2001; Waddock and Graves 1997; Russo and Fouts 1997). Nevertheless, an increasing number of studies reported that the social responsible behaviors should be used strategically (Husted and de Jesus Salazar 2006). Others argued that social responsibility offers opportunities for market differentiation, as it could be a source of competitive advantage (Russo and Fouts 1997).

Donaldson and Preston (1995) maintained that social responsibility is not fully driven by commercial factors. Their altruistic social responsibility perspective (or intrinsic stakeholder commitment) approach assumed that organizations have a normative (moral) commitment to advance their stakeholders' interests. Similarly, Castka and Balzarova (2008a) have proposed an exhaustive list of social responsibility predictors that were drawn from three perspectives: strategic, altruistic and coercive prior to the formulation of ISO 26000. They listed ten propositions in relation to social responsibility orientation of organizations or networks, differences in regulatory systems, and the role of governments and national environments.

One of the mechanisms that led to the development of the social responsibility agenda is a pressure of different groups of activists, consumers and non-governmental organizations. For instance, stakeholders may exert pressure over organizations to adopt social and environmental practices that exceed the minimum requirements that are mandated by legislation and regulation (Christmann and Taylor 2004; Corbett and Kirsch 2001). Nevertheless, there may be other stakeholders who could generate new societal expectations and consequently lead to new business practices. In fact, it is a very common practice amongst multinational supply chains to use well established codes of conducts that are imposed on others by the most powerful players (Castka and Balzarova 2008a).

Organizations ought to consider which aspects of social responsibility to invest in (McWilliams and Siegel 2001). Their social responsibility can include internal aspects (i.e. physical environment, working conditions, communication and transparency parameters) as well as external aspects (community relations, supplier relations, shareholder relations (Kok et al. 2001). McWilliams and Siegel (2001) held that there is an ideal level of CSR that managers can determine via cost–benefit analyses.

4 A Cost-Benefit Analysis of ISO 26000

4.1 Costs

When the standards are enforced, industry stakeholders need to comply with their requirements. They may be averse toward the new standards as they could assume that they could create trade barriers. This may inevitably lead to an increase in their cost of production as they strive to absorb the cost of compliance (Delmas 2002). Notwithstanding, when introducing new standards, the following external audits could reveal regulatory non-compliance among the adopting organizations (Delmas

2002). As a result, the industries' implementation of a new standard such as ISO 26000 could be time consuming, because it requires holistic adaptations to change throughout the practicing businesses.

Previous research demonstrated that some stakeholders have used the standardization to enhance their credibility, learning, and legitimacy (Boström and Hallström 2010). However, the development of ISO 26000 involved lengthy multi-stakeholder corroborations that did not necessarily ensure legitimacy or guarantee that the standard could be considered as an enforceable instrument for industry participants. Balzarova and Castka (2012) and Marques (2012) also pointed out that the scope of the ISO 26000 standard was unclear as the actual implications for social and environmental improvement were still unknown. Moreover, the standardization of social responsibility has also been criticized for being costly and thereby difficult to implement, especially among the smaller companies (Toppinen et al. 2015).

Rasche and Kell (2010) admitted that the responsibility standards can never be a complete solution to the many social and environmental problems, as their inherent limitations need to be recognized. Multiple-stakeholders have actively contributed to the development of ISO 26000, yet certain pre-standardization preparations may have created boundaries that have restricted the stakeholders' influence. Suchman (1995) described the pre-standardization phase as an effort which embedded new structures and practices into already legitimate institutions. During the pre-standardization discussions among stakeholders there were differing opinions, and not enough consensus over certification (Mueckenberger and Jastram 2010). Other authors declared that the certification of standards does not necessarily lead to improved performance (Aravind and Christmann 2011; King et al. 2005). Instead, a certificate might be used for greenwashing purposes to cover organizational conduct.

Balzarova and Castka (2012) held that some stakeholders could have been in a position to leverage their arguments during the pre-standardization arrangements. They made reference to the non-certification issue and its potential role in the standard's reinforcement. This is a contentious subject which continuously re-emerges during the stakeholders' meetings. Throughout the planning process, the idea of certifiable standards was abandoned. Previous research has indicated that a certificate can be an initial motive for using a standard (Beck and Walgenbach 2005). Debatably, in its present form, ISO 26000 may not be suitable for certification purposes since it does not follow a classical plan–do–check–act–type management system approach like, for example, ISO 14001 (Hahn 2013). Moreover, since ISO 26000 is not certifiable, some businesses could omit certain social responsibility activities if including them could lead them to release public information. The organizations that are in complete compliance with ISO 26000 may be required to publish vital information and make it readily accessible to stakeholders (Balzarova and Castka 2012). This contentious issue could cause some organizations to not fully conform themselves to the ISO guidance as they may decide to avoid disclosing their sensitive information to competitors. Thus, ISO 26000 might not offer sufficient incentives, especially for those companies which could potentially reap the greatest benefits from using it.

Hahn (2013) suggested that the ISO 26000 standard provides organizations with valuable predictability regarding their social license to operate, which is often expected from businesses by stakeholders. The author contended that each organization should recognize and address those specific areas in ISO 26000 that may be the most relevant to their own field. Ávila et al. (2013) survey found that ISO 26000's themes were under-represented, particularly those involving labor practices and the environment. Their study investigated core subjects including organizational governance and consumer issues. The authors posited that the organizations who were supposedly following ISO 26000, have often faced difficulties in incorporating the social responsibility throughout all organizational mechanisms, processes, organizational identity and decisions. Ávila et al. (2013) argued that the businesses' unsatisfactory engagement with consumer issues was even more serious as they are the stakeholders that justify the existence of organizations. It may appear that Ávila et al.'s (2013) research participants were only concerned with their corporate image (as they were supposedly implementing the social responsibility concept and its premises). Evidently, these firms were less interested in undertaking necessary actions to ensure implementation and compliance with ISO 26000.

Toppinen et al. (2015) *held* that ISO 26000 may not bring much added value to the sustainability frontrunners who had a track record in social and environmental responsibility. They also claimed that the social responsibility standard was not sufficiently detailed to incorporate forest-sector-specific issues. This reasoning also resonated with other conceptual literature that questioned whether the extant standards were contributing (or not) to the sustainable development at the national or global scale (Balzarova and Castka 2012; Hahn 2013).

4.2 Benefits

ISO 26000 has been characterized as a 'significant breakthrough innovation'(Webb 2012) and as an evolutionary step in standard innovation (Hahn 2013) because it is suitable for organizations of all sizes and sectors, and because it has unique features regarding authority and legitimacy. The standard's guideline describes social responsibility as the actions a firm takes to contribute to 'sustainable development' (Perez-Baltres et al. 2012). Hahn (2013) suggested that ISO 26000 offers specific guidance on many facets of CSR as it helps responsible businesses in their internal and external assessments and evaluations. Furthermore, when the organizations adopt ISO 26000 they could signal their social responsibility credentials and qualities to their marketplace stakeholders (Graffin and Ward 2010). They also reduce information asymmetries among supply chain partners (King et al. 2005).

ISO 26000 provides a unilateral understanding of social responsibility across the globe. It acknowledges that 'social responsibility should be an integral part of the businesses' core strategy. The definition of the standard's core subjects is a major achievement. A foundational document such as ISO 26000 is an important step in the right direction as it has broadly improved the social responsibility and sustainability agenda. Of course, there are a wide array of social responsibility issues that are addressed in ISO 26000 (as specified in Table 1). Even though the standard aims to unify and standardize social responsibility practices, it also acknowledges that each organization has a responsibility to recognize and address those areas that are relevant to its business (Hahn 2013). Notwithstanding, there are different industries, organizational settings, regional or cultural circumstances that will surely affect how every company implements the standards' responsible behaviors.

The ISO 26000 was developed through a participatory multi-stakeholder process with an emphasis on participatory decision making and democracy (Hahn and Weidtmann 2016). The ISO 26000 standard provides guidance on the integration of social responsibility into management processes, as well as on the principles on environmental responsibility and stakeholder engagement. The list of seven core subjects represents the most essential areas of SR that an organization should take into consideration in order to maximize its contribution to sustainable development.

ISO 26000 also provides guidance on stakeholder identification and engagement, assists in improving SR communications and helps to integrate responsible business practices into strategies, systems and processes. According to (Castka and Balzarova 2008a, p. 276), 'ISO 26000 aims to assist organizations and their network in addressing their social responsibilities and providing practical guidance related to operationalizing CSR, identifying and engaging with stakeholders and enhancing credibility of reports and claims made about CSR'. The standard has potential to capture the context-specific nature of social responsibility.

Corporate culture is an important driver of socially responsible activities, in which CEOs play a key role in giving their face and voice to the corporate sustainability agenda (Waldman et al. 2006; Caprar and Neville 2012). Enhancing corporate culture for sustainability has the potential to achieve a competitive advantage and improve the long term financial performance (Eccles et al. 2012) via the development of valuable, rare and non-imitable organizational resources and capabilities (Barney 1986). Eccles et al. (2012) analyzed the financial performance of firms with either high or low sustainability orientation. They found that firms with a high sustainability orientation were associated with distinct governance mechanisms for sustainability, longer time horizons, deeper stakeholder engagement and greater attention to and disclosure of non-financial measures. Adoption of sustainability standards, such as ISO 26000, can also be interpreted as a signal of a responsible corporate culture (Waldman et al. 2006). Building social responsibility systemically into strategy may thus help the companies to meet the interests of stakeholders and society as it is being suggested in the following case study.

5 Marks and Spencer Champions Sustainable Business with ISO 26000

An easy to read ISO (2015) case study features Marks & Spencer (M&S) as a leading UK-based multinational retailer that is currently following ISO 26000 guidance. The company that was founded in 1884 employs more than 85,000 employees around the globe. It has its headquarters in the City of Westminster, London. With its branches in 54 countries, the retailer has an annual turnover of over £10 billion and a net profit of £458 million (ISO 2015). M&S is listed on the London Stock Exchange and is a constituent of the FTSE 100 index. This British retailer specializes in selling clothing, homeware and luxury food products.

> In its quest to become the world's most sustainable retailer, Marks & Spencer has set a high standard for environmental sustainability and is influencing disruptive change throughout the industry.
>
> ...In 2006, M&S took a long, hard look at its ethical and sustainable activities and set out to move them up a gear, while gathering shoppers' support behind them. Enter "Plan A", which has become the core of the company's business strategy—with "overwhelmingly positive" results, according to the brand itself.
>
> For all its successes, Plan A still needed to go one step further to win customers over. Shoppers, assailed by headlines over the source of their products, like the idea that they don't have to worry if their strawberries are ethically grown or workers are treated well in factories supplying M&S. So in 2013, the company turned to ISO 26000 for guidance on how to promote respect and fair practices through its supply chain.

5.1 No Plan B

M&S turns to ISO 26000 to promote fair practices through its supply chain.Working with suppliers in 70 countries and some two million workers across 2000 factories and 20,000 farms, M&S is most definitely "supply chain dependent". This means that being a fair partner is central to the retailer's strategy, and ethical behaviour and environmental standards are essential to the future of the business.

The company wanted to go beyond the expectations of employees, customers and stakeholders by collaborating with suppliers to create a fair workplace and step up environmental performance. In 2006, it decided to put sustainability at the heart of its business, from sourcing products to relationships with suppliers, shoppers and the wider world. Behind the initiative is the goal that, by 2020, all M&S products will have at least one sustainable element—an ambitious plan demanding concrete targets, robust measurement and transparent reporting.

To succeed, it would have to change the attitudes of every one of its 25 million shoppers. In 2006, its first campaign, "Look Behind the Label", drew attention to the various ethical and environmentally friendly initiatives adopted by the company in the production and sourcing of its products. This earned them widespread approval, particularly among opinion formers and the ethically-aware.

But preaching to the converted was not enough. They needed to get people to actually "do" something. Hence the next stage was about engaging commitment. And so "Plan A" was born, launching in 2007, with 100 commitments in five pillars—climate change, waste, sustainable raw materials, fair partnership and health—to dramatically increase its environmental sustainability across the entire value chain.

Backed by an extensive audit scheme, Plan A—thus named by then CEO, Stuart Rose, on the premise that there was no Plan B—has brought fundamental changes in the way M&S does business with suppliers as well as helping it to maintain high levels of customer trust and employee engagement.

5.2 A Fair Partner

Some two million workers across 2000 factories benefit from M&S's ethical behaviour.

The brand received many accolades, yet Plan A still fell short of making the public see "green" as the desirable option. This was not helped by issues of food and labour sourcing making headlines around that time, such as the horsemeat scandal or the Rana Plaza disaster in 2013, where a garment manufacture in Bangladesh collapsed killing over a thousand workers. Although M&S had no link to these tragedies, they still posed uncomfortable questions about the human cost of affordable fashion and food.

As a major garment retailer, M&S faces challenges in responsibly sourcing materials from its suppliers in South Asia, mainly from India, Sri Lanka and Bangladesh. Yet, suppliers from these countries account for close to 80% of its footprint. What's more, the company regards the achievement of human potential as the foundation of environmental and social sustainability for present and future generations. Hence ISO 26000, Guidance on social responsibility, became their tool of choice to support the ambitious goals of Plan A. After careful consideration, ten of the biggest clothing manufacturers for M&S in India, Sri Lanka and Bangladesh—that account for 60% of the supplies—were invited to integrate the standard's principles of social and environmental sustainability into their business strategy. Of the ten, nine agreed to participate on a voluntary basis, bearing part of the cost for the programme themselves.

5.3 Out for the Long Haul

By introducing ISO 26000 to its largest suppliers, M&S ensures their allegiance to more ethical standards.

Work kicked off with a three-day capacity-building workshop in New Delhi for the nine top suppliers, where company CEO, Mark Bolland, introduced the concept, highlighting its importance for the successful implementation of Plan A. Each company dispatched two representatives who received information and training on ISO 26000, the expectations of the project, cost approximation and sundry information.

The scheme involved a gap analysis using a self-evaluation tool, an assessment by CSR Company International, an action plan based on the principles of ISO 26000, and a scorecard for implementing the agreed actions in the day-by-day running of the factories from management level to the "shop floor".

Nine-months after the first workshop, an audit of the participating companies was undertaken to check that every aspect was being applied consistently and correctly. If results were positive, the supplier would be granted the M&S "Golden Status Supplier", whereupon regular audits could be dropped in favour of occasional random checks.

5.4 Leading with Others

Keeping the communication channels open between M&S and its suppliers.

As a multinational retailer, M&S relies heavily on its suppliers for quality source materials. This means it is responsible for tracking "till the last mile" of the supply chain and checking the source of the raw materials and the labour conditions prevailing in its suppliers' organization.

By introducing ISO 26000 to its largest suppliers, M&S ensures their allegiance to more ethical standards. And by adopting ISO 26000 voluntarily, suppliers undertake to conduct their business in a more transparent and accountable manner, which will help them fit in the M&S sustainability framework set out under Plan A.

The truth is, supply-chain complexity cannot be handled only with audits; it must be based on mutual trust between supplier and buyer. To borrow Bolland's own words, the ability to "lead with others"—by treating suppliers as partners and setting clear limits—is key to delivering sector-wide change.

Lastly, the project must be championed by senior management to cement the commitment of partners and employees and impel its initiatives across the organization, driving long-term benefits.

5.5 The Onward Journey

Plan A made demonstrable business sense. With the above-mentioned strategy, M&S has 60% of its supplier purchases covered under its sustainability management framework. Had it tried to cover almost 100% of its suppliers, the sheer complexity of the project would have certainly worked against it. M&S is conducting a trial with its nine most important suppliers. Based on the success of the project, further partners may be included.

Seven years on, if there is one overarching lesson to be learned from Plan A and the journey into social responsibility, it is one of humility. Even multinationals cannot change the world alone; they need to work with their partners. Now, with its suppliers firmly behind it, M&S is ready for the next leg of its journey: Plan A 2020" (ISO 2015).

6 Implications and Conclusions

This case study suggests that globalization phenomenon is strongly shaping the businesses' behaviors across different markets. Many companies are already sourcing materials or finished items from other countries in order to reduce their manufacturing and distribution costs. Consequentially, the procurement of products may result in perceived shortcomings in the companies' behaviors. The developing countries are often characterized by poor working conditions, weak regulatory compliance procedures, and corruption among other issues. For these reasons, many stakeholders, particularly consumers hailing from the most advanced economies are increasingly inquiring on the larger corporations' responsible behaviors. In this light, this chapter has shown that social responsibility topic is an extremely broad topic as evidenced in ISO 26000 core subjects and issues; including organizational governance, human rights, labor practices, the environment, fair operating practices, consumer issues and community involvement and development.

The standard for social responsibility is an important instrument that could guide different organizations in their sustainability path. Therefore, this contribution has analyzed the costs and benefits of the social responsibility standard. Indeed, ISO 26000's recommendations are relevant and flexible to today's businesses, as they address wide array of societal and environmental issues in diverse contexts. Apart from adding academic knowledge to existent contributions on the international diffusion of ISO 26000, this chapter has also presented a descriptive case study of a multinational retail corporation. It has built on Castka and Balzarova's (2008a) work as it reported how developed and developing countries are dealing with ISO 26000 (following the standard's inception in 2010). Evidently, this non certifiable standard supports businesses on different aspects of their social responsibility, environmental sustainability and stakeholder engagement. This chapter suggested that ISO 26000 could be internally focused (for instance investigation of implementation issues within the companies e.g. safeguarding the conditions of employment, et cetera) or externally focused (e.g. social responsibility toward marketplace stakeholders, including supply chains and industries; whilst considering other stakeholder influences, including the government, labor and consumer organizations, media, NGOs et cetera).

Nevertheless, this research has offered other practical implications for policy makers, standards setting organizations, and participating stakeholders. The critical evaluation of the ISO 26000 has revealed that this broad, non-certifiable standard may not necessarily lead to the development of an enforceable instrument. Hence, policy makers should be cautious in supporting such standards because the role of such a comprehensive standard is still somewhat unclear.

7 Limitations and Suggestions for Future Research

This research focused on ISO's social responsibility standard. It drew relevant conclusions about the multiple stakeholders' influence and contribution to its development. Yet, it also reported about the costs and benefits of implementing this standard's guidelines, from the organizations' point of view. However, further inquiry is necessary in order to draw broader conclusions about how organizations are following this guiding standard as they try to satisfy their numerous stakeholder claims.

Future research could explore specific stakeholders' views on the implementation of ISO 26000. In the past there were many studies that have focused on establishing a relationship between corporate social and financial performance. Academia could easily adopt ISO's standard as a measure for social responsibility in quantitative studies. Moreover, qualitative studies could clarify and explain in detail how organizations are resorting to the core subjects and issues pertaining to ISO 26000 guidelines. Hence, further studies are suggested to map the organizations' progress and advancement in the implementation and monitoring of the standard's initiatives.

References

Aravind, D., & Christmann, P. (2011). Decoupling of standard implementation from certification: Does quality of ISO 14001 implementation affect facilities' environmental performance? *Business Ethics Quarterly, 21*(01), 73–102.

Arzova, S. B. (2009). Turkey: CSR in practice. In *Global practices of corporate social responsibility* (pp. 373–391). Heidelberg: Springer.

Ávila, L. V., Hoffmann, C., Corrêa, A. C., Rosa Gama Madruga, L. R., Schuch Júnior, V. F., & Júnior, S., et al. (2013). Social responsibility initiatives using ISO 26000: An analysis from Brazil. *Environmental Quality Management, 23*(2), 15–30.

Balzarova, M. A., & Castka, P. (2012). Stakeholders' influence and contribution to social standards development: The case of multiple stakeholder approach to ISO 26000 development. *Journal of Business Ethics, 111*(2), 265–279.

Barney, J. B. (1986). Organizational culture: Can it be a source of sustained competitive advantage? *Academy of Management Review, 11*(3), 656–665.

Beck, N., & Walgenbach, P. (2005). Technical efficiency or adaptation to institutionalized expectations? The adoption of ISO 9000 standards in the German mechanical engineering industry. *Organization Studies, 26*(6), 841–866.

Berman, S. L., Wicks, A. C., Kotha, S., & Jones, T. M. (1999). Does stakeholder orientation matter? The relationship between stakeholder management models and firm financial performance. *Academy of Management Journal, 42*(5), 488–506.

Boström, M., & Hallström, K. T. (2010). NGO power in global social and environmental standard-setting. *Global Environmental Politics, 10*(4), 36–59.
Bowers, D. (2006). Making social responsibility the standard. *Quality Progress, 39*, 35–38.
Brunsson, N., & Jacobsson, B. (2000). The contemporary expansion of standardization. *A World of Standards, 1*, 1–17.
Camilleri, M. A. (2012). *Creating shared value through strategic CSR in tourism*. PhD thesis, University of Edinburgh. Retrieved July 1, 2016, from https://www.era.lib.ed.ac.uk/bitstream/handle/1842/6564/Camilleri2012.pdf?sequence=1
Caprar, D. V., & Neville, B. A. (2012). "Norming" and "conforming": Integrating cultural and institutional explanations for sustainability adoption in business. *Journal of Business Ethics, 110*(2), 231–245.
Castka, P., & Balzarova, M. A. (2008a). ISO 26000 and supply chains—On the diffusion of the social responsibility standard. *International Journal of Production Economics, 111*(2), 274–286.
Castka, P., & Balzarova, M. A. (2008b). The impact of ISO 9000 and ISO 14000 on standardisation of social responsibility—An inside perspective. *International Journal of Production Economics, 113*(1), 74–87.
Christmann, P., & Taylor, G. (2004). Environmental self-regulation in the global economy: The role of firm capabilities. *Research in Global Strategic Management, 9*, 119–145.
Claasen, C., & Roloff, J. (2012). The link between responsibility and legitimacy: The case of De Beers in Namibia. *Journal of Business Ethics, 107*(3), 379–398.
Conroy, M. E. (2007). *Branded! How the certification revolution is transforming global corporations*. Gabriola Island: New Society Publishers.
Corbett, C. J., & Kirsch, D. A. (2001). International diffusion of ISO 14000 certification. *Production and Operations Management, 10*(3), 327–342.
Delmas, M. A. (2002). The diffusion of environmental management standards in Europe and the United States: An institutional perspective. *Policy Sciences, 35*, 91–119.
Donaldson, T., & Preston, L. E. (1995). The stakeholder theory of the corporation: Concepts, evidence, and implications. *Academy of Management Review, 20*(1), 65–91.
Eccles, R. G., Ioannou, I., & Serafeim, G. (2012). *The impact of a corporate culture of sustainability on corporate behavior and performance* (no. W17950). National Bureau of Economic Research.
Figge, F., Hahn, T., Schaltegger, S., & Wagner, M. (2002). The sustainability balanced scorecard–linking sustainability management to business strategy. *Business Strategy and the Environment, 11*(5), 269–284.
Fransen, L. W., & Kolk, A. (2007). Global rule-setting for business: A critical analysis of multi-stakeholder standards. *Organization, 14*(5), 667–684.
Graffin, S. D., & Ward, A. J. (2010). Certifications and reputation: Determining the standard of desirability amidst uncertainty. *Organization Science, 21*(2), 331–346.
Hahn, R. (2013). ISO 26000 and the standardization of strategic management processes for sustainability and corporate social responsibility. *Business Strategy and the Environment, 22*(7), 442–455.
Hahn, R., & Weidtmann, C. (2016). Transnational governance, deliberative democracy, and the legitimacy of ISO 26000 analyzing the case of a global multistakeholder process. *Business & Society, 55*(1), 90–129.
Husted, B. W., & de Jesus Salazar, J. (2006). Taking Friedman seriously: Maximizing profits and social performance. *Journal of Management Studies, 43*(1), 75–91.
Idowu, S. O., & Leal Filho, W. (2009). *Global practices of corporate social responsibility*. Berlin: Springer.

ISO. (2004). *ISO/TMB policy and principles statement global relevance of ISO technical work and publications*. TMB/SC/GR 2004-06-30. Retrieved June 10, 2016, from http://www.iso.org/iso/global_relevance.pdf
ISO. (2014). *ISO 26000: Guidance on social responsibility*. Retrieved June 8, 2016, from http://www.iso.org/iso/discovering_iso_26000.pdf
ISO. (2015). *M&S champions sustainable business with ISO 26000*. Retrieved June 12, 2016, from http://www.iso.org/iso/home/news_index/news_archive/news.htm?refid=Ref1997
King, A. A., Lenox, L. J., & Terlaak, A. (2005). The strategic use of decentralized institutions: Exploring certification with the ISO 14001 management standard. *Academy of Management Journal, 48*(6), 1091.
Kok, P., Van der Wiele, T., McKenna, R., & Brown, A. (2001). A corporate social responsibility audit within a quality management framework. *Journal of Business Ethics, 31*, 285–297.
Marques, J. (2012). A global SR standard: Good, or too good to be true? *The Journal for Quality and Participation, 34*(4), 29.
McWilliams, A., & Siegel, D. (2001). Corporate social responsibility: A theory of the firm perspective. *Academy of Management Review, 26*(1), 117–127.
Moratis, L. (2015). The credibility of corporate CSR claims: A taxonomy based on ISO 26000 and a research agenda. *Total Quality Management & Business Excellence, 281*(2), 147–158.
Mueckenberger, U., & Jastram, S. (2010). Transnational norm-building networks and the legitimacy of corporate social responsibility standards. *Journal of Business Ethics, 97*(2), 223–239.
Orlitzky, M., Schmidt, F. L., & Rynes, S. L. (2003). Corporate social and financial performance: A meta-analysis. *Organization Studies, 24*(3), 403–441.
Perez-Baltres, L., Doh, J., Miller, V., & Pisani, M. (2012). Stakeholder pressures as determinants of CSR strategic choice: Why do firms choose symbolic versus substantive self-regulatory codes of conduct? *Journal of Business Ethics, 110*, 157–172.
Rangan, S., Samii, R., & Van Wassenhove, L. N. (2006). Constructive partnerships: When alliances between private firms and public actors can enable creative strategies. *Academy of Management Review, 31*(3), 738–751.
Rasche, A., & Kell, G. (2010). *The United Nations global compact: Achievements, trends and challenges*. Cambridge: Cambridge University Press.
Russo, M. V., & Fouts, P. A. (1997). A resource-based perspective on corporate environmental performance and profitability. *Academy of Management Journal, 40*(3), 534–559.
Schwartz, B., & Tilling, K. (2009). 'ISO-lating' corporate social responsibility in the organizational context: A dissenting interpretation of ISO 26000. *Corporate Social Responsibility and Environmental Management, 16*(5), 289.
Suchman, M. C. (1995). Managing legitimacy: Strategic and institutional approaches. *Academy of Management Review, 20*, 571–610.
Toppinen, A., Virtanen, A., Mayer, A., & Tuppura, A. (2015). Standardizing social responsibility via ISO 26000: Empirical insights from the Forest industry. *Sustainable Development, 23*(3), 153–166.
Waddock, S. A., & Graves, S. B. (1997). The corporate social performance-financial performance link. *Strategic Management Journal, 18*, 303–319.
Waldman, D. A., de Luque, M. S., Washburn, N., House, R. J., Adetoun, B., Barrasa, A., & Dorfman, P. (2006). Cultural and leadership predictors of corporate social responsibility values of top management: A GLOBE study of 15 countries. *Journal of International Business Studies, 37*(6), 823–837.
Webb, K. (2012). *ISO 26000: Bridging the public/private divide in transnational business governance interactions*. Osgoode CLPE Research Paper, 21.

Mark Anthony Camilleri, PhD is a resident academic in the Department of Corporate Communication. He lectures in an international masters programme run by the University of Malta in collaboration with King's College, University of London. Mark specialises in strategic management, marketing, research and evaluation. He successfully finalised his PhD (Management) in 3 years time at the University of Edinburgh in Scotland—where he was also nominated for his "Excellence in Teaching". During the past years, Mark taught business subjects at under-graduate, vocational and post-graduate levels in Hong Kong, Malta and the UK.
Dr. Camilleri is a member in the Global Reporting Initiative (GRI)'s Stakeholder Council, where he is representing Europe and Asia's CIS region. He is a member on the editorial board of Springer's International Journal of Corporate Social Responsibility and Inderscience's International Journal of Comparative Management and the International Journal of Responsible Management in Emerging Economies. Mark is also a member of the academic advisory committee in the Global Corporate Governance Institute (USA). He has published his research in reputable peer-reviewed journals. He is a frequent speaker and reviewer at the British Academy of Management and in the American Marketing Association's (AMA) annual gatherings.
Dr. Camilleri's first book, is entitled; "Creating Shared Value through Strategic CSR in Tourism" (2013). This year he published two Springer textbooks; "Corporate Sustainability, Social Responsibility and Environmental Management: An Introduction to Theory and Practice with Case Studies" (2017), and 'Travel Marketing, Tourism Economics and the Airline Product'. Moreover, he edited "CSR 2.0 and the New Era of Corporate Citizenship" (that was indexed in Scopus).

Aspects of ISO 26000 Related to Transaction and Environmental Costs

Ninel Nesheva-Kiosseva

1 Introduction

ISO 26000, Guidance on Social Responsibility, has been set up by the Social Responsibility Working Group at the Technical Council of ISO. Its text has been approved by many countries and 42 public and private sector organizations and was published in 2010 (Economic Benefits of Standards 2013, p. 304). Standard IC CSR-08260008000 "Social Responsibility Requirements" was adopted the following year, complementing it. ISO 26000 covers all corporate social responsibility fields. It is built around "Seven Principles":

1. Accountability;
2. Transparency and disclosure;
3. Ethical behaviour;
4. Respect for the stakeholders' interests—Stakeholder engagement;
5. Respect for rule of law;
6. Respect for international norms of behaviour;
7. Respect for Human rights (ISO 26000 2010, pp. 10–14).

ISO 26000 proclaimed "Seven core subjects and their related issues" and "Stakeholder engagement". Core subjects are: Organizational governance; Human rights; Labor practices; Environment; Fair operating practices; Consumer issues; Community involvement and development (ISO 26000 2010, pp. 19–69).

The seven core subjects address the following five problem areas: Economy; Health and safety; Value chain; Gender equity; Communication with stakeholders (ISO 26000 2010, pp. 19–69).

N. Nesheva-Kiosseva (✉)
New University of Bulgaria, Sofia, Bulgaria
e-mail: nkioseva@nbu.bg

S. O. Idowu et al. (eds.), *ISO 26000 - A Standardized View on Corporate Social Responsibility*, CSR, Sustainability, Ethics & Governance,
https://doi.org/10.1007/978-3-319-92651-3_11

2 Research Questions

The article aims to give an answer to the questions:

1. What are opportunities, following ISO 26000, to clarifying, separating and analyzing transaction and environmental costs?
2. Which transaction and environmental costs can be identified and analyzed on the basis of the ISO 26000 methodology?
3. Does the practical use of ISO 26000 lead to the optimization of transaction and environmental costs in business and does it improve its economic and financial performance?

3 For Profit or for Loss

3.1 Criticism and Acceptance

Despite the fact that ISO 26000 speaks for itself, there are different analyses of it.

In general, opinions of ISO 26000 are divided into two groups: the group rejecting it, and the group supporting it.

The critics of ISO 26000 ideology and its application are of the neo-liberal economists' circle. Even before the appearance of ISO 26000 as a structuring concept, the idea of corporate social responsibility was met critically. The leading figure of neo-liberal economics in the 1970s, Milton Friedman published in the New York Times an article in which he argued the complete unacceptance of "corporate social responsibility" from the point of view of free business, free market, and of course—profit. In an almost ironic fashion, Friedman concedes that the doctrine of CSR is absurd in terms of "rules of the game" in free capitalism, ending with its unambiguous rejection: "*There is one and only one social responsibility of business—to use its resources and engage in activities designed to increase its profits so long as it remains within the rules of the game, that is, engages in open and free competition without deception or fraud*" (Friedman 1970).

After the publication of ISO 26000, neo-liberal economists, in line with Freidman's criticism, point to several sets of arguments against the adoption of the standard.

The first group of arguments is concerning its impact on the free market economic system. They believe that ISO 26000:

1. Infringes the rules of free competition and free trade;
2. Therefore undermines the market;
3. Hinders the natural development of market economic relations.

The second group of liberal arguments concerns its "noncompulsory" aspect. There is, in fact, from their point of view, administrative pressure on companies to accept the otherwise voluntary ISO 26000. This hidden administrative pressure, forcing organizations to adopt the Standard's guidelines, would, in their view, have the effect of the regulatory authorities interfering in the work of companies. In this way, "Free Entrepreneurship" is also under attack by the Standard. The ISO 26000 is seen as a plan for "global business regulation" (Roberts 2010).

Despite Friedman's argumentation, followed by the criticism of the neo-liberal economists, the Standard begins to be considered as "national" and becomes part of the management of more and more firms. "*ISO 26000 has been adopted as a national standard in over 80 countries (and counting) and its text is available in some 22 languages. It is also referenced in more than 3000 academic papers, 50 books and numerous doctorates, and is used by organizations of all shapes and sizes including Petrobras, Air France, British Telecom, NEC, NovoNordisk and Marks & Spencer, to name a few*" (Naden 2017).

ISO 26000 quickly becomes a popular "external" managerial standard that covers all areas and relationships of business activity. The International Organization for Standardization has the ambition to see its standards used everywhere, including ISO 26000. In practice it appears that ISO 26000 not only does not diminish or liquidate the profit of companies that voluntarily accept it as the guiding rule of their activity, but, on the contrary, such companies are doing well economically. Nothing goes wrong with their ability to generate profits.

With the reinforcement of the scientific paradigm of neo-institutionalism, increasing attention is being devoted not to the production function but to the management function. In practice, it turns out that when a company comes out of its "black box", embracing the ISO 26000 philosophy this is a means of achieving its economic and financial sustainability.

4 Transaction and Environmental Costs in ISO 26000 'Practical Philosophy'

The methodology of ISO 26000, as well as the methodology of all standards in the ISO group, is based on a Value-chain model (Economic Benefits of Standards, International Case Studies 2013, vol. 2, p. 14). Using this model, the value added generated during every stage of the production process, distribution and sales can be tracked and reported. By using accounting tools this approach also allows the tracking of costs and their classification according to the information needs of the management and stakeholders. Here we will specifically take a look at the opportunities for separating and exposing the transaction and environmental costs of a company based on the ISO 26000 methodology.

The easiest way to reduce costs is to reduce OPEX.[1] OPEX contains a substantial amount of transaction-related costs. Conventional financial accounting does not take

---

[1]OPEX—Operating expenses. Operating expenses are the company's daily expenses for the organization of sales, administration, R&D, expenses for employees etc. Operating expenses are

into account and does not show many of the existing operating costs that are important to the company's performance—effectiveness, cost of output etc.—the transaction costs.

Conventional accounting also does not separate environmental costs. Environmental costs can be transformation costs arising during production, as well as investment costs for acquiring assets with an environmental purpose, fines, fees, etc., external costs, which can also be characterized as recycling, cleaning up pollution, etc., some of which are OPEX.

The separation of environmental and transaction costs makes these clear to the management enabling it to focus its attention on them. Some of these costs are visible, like litigation costs, but many of them are invisible.

By implementing ISO 26000 the management focuses its attention on these costs as well as on their individual analysis. Appropriate company policy aimed at optimizing these costs, can bring about successful economic performance.

Broadly speaking, transaction costs are the costs that a firm makes in dealing with other market players and, in a narrower aspect, the costs that the firm makes between the stages of the production process. The latter are costs that do not add up value-related to time, such as stops, control, monitoring.

High transaction costs per unit of transaction lead to the company's lack of competitiveness and its dropping out of the market, hindering the development of the markets and hence competition.

One of the benefits that organizations receive through the implementation of ISO 26000 is the reduction of transaction costs—in a broad and narrow sense.

The Standard's methodology has as its primary objective the reduction of transaction costs, related to procurement and relations with suppliers (ISO 26000 2010, p. 17).

The Standard focuses on reducing transaction costs in the path of so-called Supplier Relationship Management—administrative costs in the relationship with suppliers, and ensuring predictability when planning them. The methodology also enables forecasting and planning of transaction costs of this type in the future. The process of highlighting these transaction costs involves the selection of suppliers, choosing new types of production and raw materials, managing the supply cycle and optimizing it. This is essential for the future sustainability of the company.

Transaction costs that can be highlighted, analyzed and optimized in the light of procurement and relations with suppliers, using ISO 26000 are:

- Costs arising from failure of suppliers to fulfill contract;
- Costs arising from negative reputation of suppliers;
- Costs relating to bad supply efficiency;
- Costs from failure to deliver or delay of delivery;

---

reported in the statement of financial performance in relation to the time—the period during which they are incurred (1 month, quarter or year).

- Costs from insufficient delivery system performance, which could have been changed more effectively, if clarity for suppliers and new supply systems were available;
- Costs from establishment of property rights in particular for the supply of goods subject to copyright and patent law;
- Informational costs;

As well as "ex post" costs like:

- Costs of monitoring suppliers and the quality of deliveries;
- Costs for monitoring the goods and resources market;
- Some costs of measuring the valuable properties of the goods and services supplied;
- Expenditure ensuing from non-compliance of contracts.

From an accounting point of view both groups of transaction costs can be significant and generally characterized as expenses (losses). This approach can help optimize reserves, prevent order failures, which also means a reduction in overfeeding losses, defaults on outstanding contracts and reputational losses. As a whole this means possibilities of avoiding losses. The application of the standard enables balancing the delivery system and reducing the negative risks.

The exiting of broader markets and increased sales at the "end of the chain" as a result of the application of the Standard also aims to reduce company transaction costs like costs for targeting customer groups, searching for customer niches, advertising, information costs, marketing research and consequently increases sales efficiency.

The reduction in transaction costs reflects on the key indicator of the competitiveness of organizations—the sale prices. As a result of the reduction in costs, the prices of goods and services can also be reduced. Increasing profit rather than reducing it is the practical result of reducing these transaction costs, which are actually losses. On the other hand, an increased profit leads to the possibility of raising salaries and investments with a company's own funds. This creates opportunities for social and environmental costs, for investments with one's own sources in assets with environmental purpose and in intellectual capital.

The "Value-chain model", according to the methodology of the Standard, is also applied to the environmental activities of the company.

Most countries have their own methodology for determining environmental costs, but there are also countries that have not developed such a methodology. The ISO 26000 setting for the "Value chain" approach and the "Seven Core Subjects" of the Standard directs companies to separate and subtract environmental costs (ISO 26000 2010, pp. 60–67, 76–77, 80–81, 103).

5 Treatment of Environmental Costs in the ISO 26000 Principles

ISO 26000 considers the two main parts of environmental corporate responsibility—Environmental protection and Cost-effective use of resources (Fig. 1).

Environmental protection creates value for the company, also in environmental or environmental-related costs. At first glance, it is strange someone to give to the companies a guidance, for making expenses when the major obligation of the company is to create a profit.

What are "environmental costs"? The easier and cleaner definition is derived from statistics, which collect data from companies' ecological statements concerning their ecological expenses. "*Environmental costs are costs connected with the actual or potential deterioration of natural assets due to economic activities. Such costs can be viewed from two different perspectives, namely as (a) costs caused, that is, costs associated with economic units actually or potentially causing environmental deterioration by their own activities or as (b) costs borne, that is, costs incurred by economic units independently of whether they have actually caused the environmental impacts*" (The OECD Glossary of Statistical Terms 2003). Accounting of environmental costs is logically equivalent with the statistics. There are differences between national accounting standards, and specifically with the environmental accounting of the various economic sectors. The *environmental expenditure "categories, reflecting the type of environmental activity (e.g., waste management and waste prevention)"* (International Federation of Accountants 2005, p. 37).

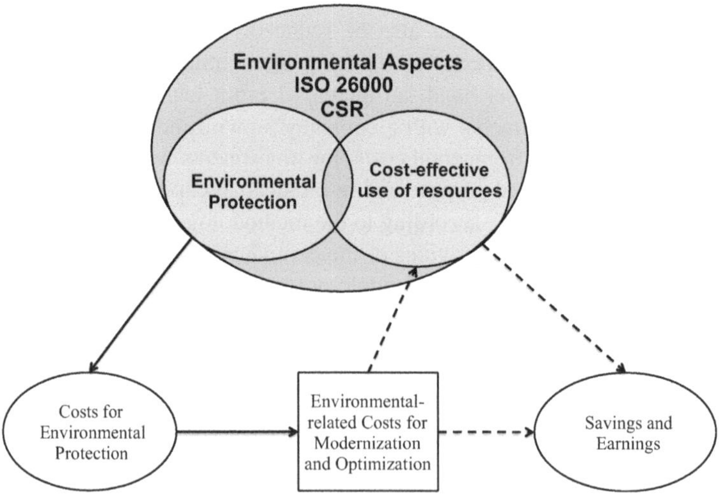

Fig. 1 Main element of environmental corporate responsibility in ISO 26000. Source: Own figure

- Costs which are used for specifically direct production, like specific direct materials and labor at specific prices and quantities;
- Costs of global and local importance for conservation, purification and/or restoration of water, air, land.
- Costs, which can or cannot be seen with the tools of conventional accounting categories reflecting visible and hidden costs ((International Federation of Accountants 2005, p. 37).

The Guidance for Environmental Management Accounting can be successfully applied when adopting Standard 26000 by the companies. In addition, it is ISO 1451: 2011 Material Flow Cost Accounting, developed from International Organization for Standardization at 2011, which provides companies with a cost-effective environmental cost management model, based on the Value—chain approach.

In general, there are six categories of environmental costs:

1. Material Costs of Product Outputs;
2. Material Costs of Non-Product Outputs (such as scrapped production that does not add value);
3. Waste and Emission Control;
4. Prevention and Other Environmental Management Costs;
5. Research and Development Costs;
6. Less Tangible Costs as the company's environmental reputation (International Federation of Accountants 2005, p. 38).

The ISO 26000 methodology combined with ISO 14051 and the Guidance for Environmental Management Accounting enables these costs, both visible and hidden, to be captured, highlighted and categorized for decision-making purposes.

This approach also helps for costs to be allocated by type beyond the standard cost-classification in accounting. All this helps to make concrete decisions in specific areas of cost management and to reduce different business risks.

The 6th part at ISO 26000, Guidance on social responsibility core subjects, 6.5.1.2 addresses the environmental costs. They are considered against four principles.

1. Environmental Responsibility;
2. Precautionary Approach;
3. Environmental Risk Management;
4. Polluter Pays (ISO 26000 2010, p. 51).

As a whole, in this part ISO 26000 gives organizations guidance on *"Compliance—Ecological Effectiveness—Strategic Position"* and the relationship between them. It is recommended to use "Due diligence" to develop a detailed and objective notion of environmental costs, savings and income from environmental protection and investing in the economic use of resources (ISO 26000 2010, pp. 12, 23, 26, 32, 34, 36–37, 40, 60, 80–82). These principles are consistent with Environmental Management Accounting, International Guidance Document (Environmental Management Accounting, p. 8).

Environmental responsibility is related to the conformity of the company's environmental policy with laws and regulations. This corresponds to the Ordo-liberal "Structural Society Concept". In the event of non-compliance with environmental regulations there is a probability for liability costs to emerge (Environmental Management Accounting, p. 49). A company that does not abide to the laws and regulations of a state or on an international level concerning environmental protection is subject to various punishments such as fines, fees, writing-off of assets and even risks to cease activities. These liability costs may severely damage the company and it may incur enormous lost due to the fact that it has not complied with ecological requirements. With the implementation and compliance of ISO 26000 the company saves grave and possibly fatal expenses of an environmentally friendly nature.

The second item "precautionary approach" (ISO 26000 2010, p. 51) is drawn up from the Principle 15 of "Rio Declaration" (The Rio Declaration 1992, p. 3). This point is in connection with the need to evaluate the effectiveness of environmental costs, even when there is no scientifically proven system for that. "*When considering measures of cost effectiveness, organizations must consider the long-term costs and benefits of that measure, not just short-term costs of the organization*" (ISO 26000 2010, p. 41).

Here ISO 26000 poses several problems:

1. On the effectiveness of environmental costs;
2. For the effectiveness of the measures taken by the organization to optimize environmental costs;
3. For the short and long-term environmental costs and benefits of the organization;
4. The danger for companies—polluters to follow-up ecologically as a result of sanctions for non-compliance with the "Polluter Pays" principle.

Solution of these four issues requires the use of in-company accounting tools for environmental costs (and revenues) as well as an approach for their forecasting and planning over a long period of time.

Environmental costs have different classifications within the different accounting systems depending on the focus on them. Not only the states, but also the different organizations may have a different classification of environmental costs and use different methods of calculating them. Like transaction costs, they can be classified as "ex ante" and "ex post" costs. More substantial attention is paid to these ex ante costs as they are cost-effective and save significant sunk costs, which are a large part of "ex-post" costs.

ISO 26000 encourages organizations to carry out environmental cost accounting and see that they are measured and managed, even in the case of "lack of full scientific certainty" to "guard" against possible environmental damage and negative financial consequences for them.

Some countries have not adopted environmental accounting standards. Nevertheless, the organization may voluntarily take action to clarify its environmental costs and the benefits of this. This is the example with the company presented in the case study below.

ISO 26000 directs practically organizations to display, manage, measure the effectiveness of their current and future environmental costs, and classify them as preventive, current and predictive.

Creating accounting records for environmental costs, especially in the absence of a nationally accepted accounting standard is a specific job, depending on the sector in which the company operates, the specificity of its production, the geographic location, the infrastructure it has or uses. For this reason, even when there is no "full scientific certainty" about the methodology and approach to be used, it is necessary, in order to achieve cost effectiveness, that the organization should start identifying and measuring these. At the same time, it must take into account how the different environmental costs it makes or is supposed to make in the future will affect its economic performance.

An organization that adopts the guidance of ISO 26000 works not only for the creation of an economic or environmental value but also for both.

Within the sphere of prevention of sunk environmental costs is the ISO 26000 recommendation for "environmental risk management". "*An organization should implement programs using a risk based and sustainability perspective to assess, avoid, reduce and mitigate environmental risks and impacts of its activities. An organization should develop and implement awareness-raising activities and emergency response procedures to reduce and mitigate environmental, health and safety impacts caused by accidents and to communicate information on environmental incidents to appropriate authorities and local communities*" (ISO 26000 2010, p. 51).

Preventive environmental costs contribute to the organizations' future benefits from the consequences of environmental incidents that may arise as a result of their operations. Ordinary insurance is not always able to cover the damage from such incidents.

Environmental incidents and systematic environmental pollution, which can cause unplanned and high expenses to businesses, are also be the subject of the next key set-up—"Polluter Pays". Once ecological impacts have been identified and their sources have been indicated, measured and reported, the organization should seek to "internalize its externalities". The organization must not transfer the costs to society, which is forced to pay these while the organization appropriates the profits. The organization must create a system of "internalization of externalities" for the sake of justice and its own social responsibility.

In view of the proper management of environmental costs and the internalization of environmental externalities, ISO 26000 recommends that the organization adopt Life Cycle Approach (LCA)—downstream/upstream—from supplier via inputs of raw materials and energy—to recycling and disposal and end-of-life of products, to reducing emissions (ISO 26000 2010, pp. 12, 30, 52). LCA is an aggregation and evaluation of the inputs, outputs and the potential environmental impacts of a product system throughout its life cycle. Several techniques have subsequently been developed in this area, including the International Organization for Standardization (ISO), ISO/TS 14048, Environmental Management—Life cycle assessment—Data, documentation format, ISO/TR 14049, Environmental management—Life cycle assessment—ISO 14041, Environmental and Environmental Management—Vocabulary, ISO 14051, Environmental management—Material flow cost accounting, ISO/TR 14062, Environmental management—Integrating environmental aspects into product design and development, etc.

The Life Cycle Approach creates conditions for cost cuts and contributes to a number of economic benefits such as reducing costs for raw materials and energy, improving resource and energy efficiency, reducing waste and re-using waste materials. Greenhouse gas reduction creates conditions for emission trading, which is another economic benefit for organizations, while enhancing environmental savings.

6 Application of ISO 26000 and Its Results (Case Study)

In practice, identifying transaction costs is not easy. Their magnitude depends on many factors such as the size of the market, taxes, etc. Environmental costs, although physically and financially measurable, also depend on specific environmental legislation and regulations, tax relief for environmental investments, the norms adopted in each municipality, the status of the area in which the organization is located etc.

We will take a brief look at the application of ISO 26000 by a private enterprise, in the absence of state incentives for conducting environmental policy and the non-separation of transaction costs in the conventional financial statement.

According to the information given from the NGO "CSR Bulgaria", nine companies operating on the territory of the country systematically apply ISO 26000 (including four Bulgarian). They are: Aurubis (metals, Germany), Solvay Sodi (industrial chemistry, Bulgaria), Globul (telecommunications, Bulgaria) Danone (Foods, France), Max Europe (Bicycles, Strollers, Bulgaria), UBB (Banks, Greece), Telus (Canada, telecommunications) and Musala Soft Ltd (Software Products and Services, Bulgaria) (Corporate Partners 2017).

Among these companies, "Musala Soft LTD" is a Bulgarian company with a history of 17 years, starting from a zero level, with continuously expanding markets and an increasing number of employers. "Musala Soft" is the first Bulgarian company adopting ISO 26000 in its practice (Musala Soft 2016, p. 5). The company is also certified by the ISO 14001 Environmental Management System and recognizes ISO 26000 as a reference document.

The company started with four employees in 2000 and has increased their number to 400 in 2017. It already has offices in Sofia, Varna, Burgas, Ruse (in Bulgaria) and Skopje (Republic of Macedonia) (Musala Soft 2017a). The Company is planning to open up new jobs and in 2020 the number of its employees will reach 1000 (Musala Soft 2017c). 10% of its total costs are directed to investment (Prizma 2017).

It is the first Bulgarian company receiving the "Software Excellence" award in the category "Small and Medium Enterprises" by the European Software Institute in 2010 (Invest Bulgaria Agency 2017). Musala Soft is a national winner at the European Business Awards for 2016/2017 in the IT sector—"Employer of the Year" (The European Business Awards 2017). The company has won the prestigious

"E-volution Awards, 2016" by Forbes Bulgaria, in the category "Going Abroad". E-volution awards are honors bestowed for the most innovative practices in the field of information technology (Musala Soft 2017b, pp. 1, 6).

All these testify to the positive development of the company and come from its adoption of ISO 26000.

"Musala Soft" follows the ISO 26000 recommended line "Compliance-Environmental Efficiency-Strategic Position", though still in a completely unexplored form.

Musala Soft presents public reports in compliance with state's legislation, environmental accountability, social responsibility directed to suppliers, customers and employees (Musala Soft 2016, pp. 18–19). Its activity in this direction helps to lower its transaction costs as discussed above.

Musala Soft reports "Significant Environmental Aspects" for: Emissions to Air, Musala Soft, pp. 14–15 and "Other Environmental Aspects" (Musala Soft 2016, pp. 18–19).

Enhancing energy efficiency is the basis for reducing production costs, improving investment opportunities and bringing about renewal of used technology.

Musala Soft reports its greenhouse emissions exclusively for strategic purposes, as there is no tax cuts in the country for companies for environmental expenses. The company cannot obtain tax relief for realized environmental investments under the Corporate Income Tax Act, as relief is provided only for foreign direct investment (Musala Soft 2016, pp. 1–20). Carbon emissions reports show that the expansion of activity has seen an increase in carbon emissions from road transport. At the same time, "Musala Soft" makes a successful effort to reduce emissions from the use of air transport, which is the largest pollutant among modes of transport.

Future revenues the company will achieve when lowering its emissions and sell verified quotas on the carbon market. This strategic activity of the company is justified by the improvement of the conditions for trading with greenhouse gases envisaged at European and world level. Following the dramatic fall in carbon prices, the European Climate Commission has planned measures for raisin and stabilizes their prices in order to create more benefits for companies that are spending money to reduce their greenhouse gas emissions (European Commission, 2017, February 16). Along with other emission mitigation measures, to reach the objectives of the Paris Accord, the Climate Commission believes that in 2020 the cost of hydrocarbons will be from \$40 to \$80, and in 2030 it will be in the range of \$50 to \$100 (from the present level of 8 \$/ton (CO2 European Emissions Allowance 2017, November 20). These measures would contribute to significant environmental revenue for smaller carbon footprint companies.

In terms of its environmental policy, the company has set up a model for accountability and disclosure of its environmental performance in compliance with key environmental reporting indicators (Table 1).

With its environmental responsibility policy, voluntarily following the guidance of ISO 26000 "Musala Soft" eliminates the risk of non-beneficial costs such as fines and penalties for air pollution (Clean Air Act, Local), for waste tax (Local Taxes and Fees Act), tax on costs related to the operation of motor vehicles used for management activity (National Income Agency of Republic of Bulgaria 2015).

Table 1 Environmental performance data, significant environmental aspects (Musala Soft 2013–2016)

Year/Subject	2013	2014	2015	2016	Trend
Electricity use (MWh/per employee)	1703	1300	1258	1235	
Emissions to air: GHG emissions generated by electricity use, tons CO_2/MWh (per employee)	1588	1212	1173	1152	
Fuel used for company car (L)	580	428	589	1736	
Fuel used for taxies	5700	5200	7000	1100	
Fuel used for air travel	4576	5175	1925	1153	
GHG emissions generated by company car (CO_2 kg)	1552	1146	1576	4646	
GHG emissions generated by use of taxies (CO_2 kg)	1354	1235	1662	2612	
GHG emissions generated by Air Travels	30995	32775	169.33	101.42	
Paper and cardboard waste generated (kg)	552	369	273	304	
Batteries waste (kg)	109.2	133.1	128	42	
Toner cartridges waste (number)	33	19	24	21	
Plastic waste (kg)	1386	1358	1480	1258	
Electronic equipment	0	0	70	75	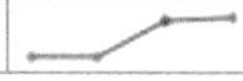

Source: Musala Soft (2014), Corporate citizenship and environmental performance report, pp. 11–13; Corporate citizenship and environmental performance report (2016, pp. 14–16)

The ISO 26000 methodology recommends EBIT[2] as a most general and representative indicator for the companies using it. Based on the data from the annual financial statements of Musala Soft, this indicator shows in Fig. 2.

The 2011 crisis is taken as the baseline year, which led to the plunge of a large number of companies.

In 2013, which is not one of the most successful for the company, Musala Soft introduces the principles of CSR in its work. The results are already visible in 2014.

Based on the Standard methodology, Musala Soft has focused on the transaction and environmental aspects of its business and, accordingly, on its transaction and environmental costs. As far as transaction costs are concerned, it complies with the recommendations of the standard, creating a Business analysis department. Through it the company applies "procurement and relations with suppliers". "*Our Business Analysis experts translate the business need into an objective, set a common, understandable language for customers and development teams, build strong relationships with all stakeholders, manage business needs throughout the whole project life cycle;*

[2]EBIT—Earnings before interest and taxes. EBIT = Revenue−Operating Expenses (OPEX).

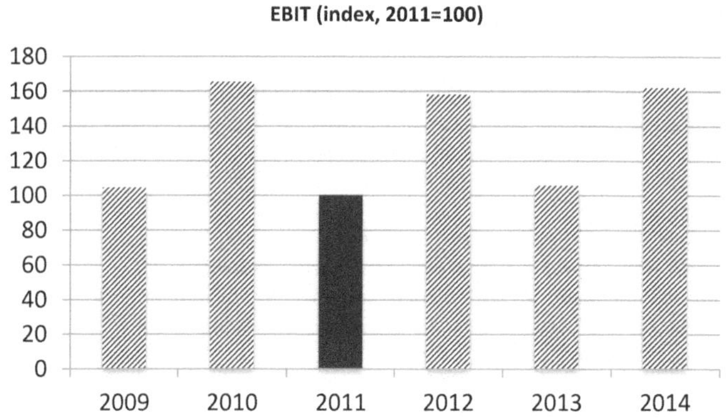

Fig. 2 EBIT, Musala Soft Ltd. 2009–2014, index, 2011 = 100. Source: Musala Soft (2017d, Annual financial reports for 2009, 2010, 2011, 2012, 2013, 2014)

analyze and identify the scope of business needs and process gaps, define and document scenarios and use cases; provide software functional and non-functional requirements analysis, document analysis, provide data & process modeling, visual modeling; document business needs and software requirements; develop prototypes and mockups. We provide a combination of business analysis, process analysis and project management in order to achieve the best results together" (Musala Soft 2017c).

7 Conclusions and Discussions

In a theoretical and practical aspect, ISO 26000 provides tools and formulates objectives for highlighting and managing transaction and environmental costs of organizations that go across their entire business.

The application of the Standard provides the possibility of creating an efficient procurement, negotiation and sales infrastructure. This infrastructure is a costly transaction sector. Simultaneously with the optimized transaction environment, the organization manages to carry out its transformative production processes sustainably by directing its analyses towards environmental costs and their special management. Appropriate business models of the implementation of ISO 26000 can to be an anti-crisis mechanism for many organizations.

Environmental accounting is a well-developed field. Environmental management and reporting standards have been developed at different levels. This is not the case with accounting and reporting of transaction costs, which are estimated as mainly in the financial enterprises. The further development of implementing ISO 26000 will possibly be in the non-financial sector, bearing in mind that this Standard will be unique in nature and content for each organization.

References

CO_2 European Emissions Allowance. (2017, November 20). Emission allowances PRICE today I Price of CO_2 European emission allowances and chart CO_2 European. *Business Insider*. Retrieved November 20, 2017, from http://markets.businessinsider.com/commodities/co2-emissionsrechte;

Corporate Partners. (2017). *CSR Bulgaria*. Retrieved November 15, 2017, from http://www.csr.bg/partners/corporate;

European Commission. (2017, February 16). *Climate action, international carbon market* [html]. EC. Retrieved from https://ec.europa.eu/clima/policies/ets/markets_en;

Friedman, M. (1970, September 13). The social responsibility of business is to increase its profits. *The New York Times Magazine*, New York. Retrieved from https://www.iso.org/files/live/sites/isoorg/files/archive/pdf/en/pub100,311.pdf

International Federation of Accountants. (2005). *Environmental management accounting, International guidance document* (p. 37). New York: IFA.

Invest Bulgaria Agency. (2017). Sectors, information technologies, Musala Soft. Retrieved October 12, 2017, from http://www.investbg.government.bg/en/sectors/successful-example-h18-46.html

International Organization for Standatization. (2010, November 01). *ISO 26000 Guidance on social responsibility* (1st ed.). Geneva: ISO.

International Organization for Standatization. (2011). *ISO 1451 Material flow cost accounting*. Geneva: ISO. Retrieved January 19, 2018, from https://www.iso.org/obp/ui/#iso:std:iso:14051:ed-1:v1:en

Musala Soft. (2017a). *Musala Soft just become 100 at Skopje, 20 Mart*. Retrieved October, 12, 2017, from http://www.musala.com/featured/musala-soft-just-became-100-at-skopje/

Musala Soft. (2017b). *CSR and environmental management policy* (pp. 1, 6). Retrieved October, 12, 2017, from http://www.musala.com/website/wp-content/uploads/2017/01/CSR-and-Environmental-Management-Policy.pdf

Musala Soft. (2017c). *Business analysis, understanding, not just analyze*. Retrieved October, 12, 2017, from http://www.musala.com/services/business-analysis/

Musala Soft. (2017d). *Annual financial reports of Musala Soft Ltd. for 2009, 2010, 2011, 2012, 2013, 2014*. Retrieved October, 12, 2017, from commercial register of Bulgaria https://public.brra.bg/checkups/default.ra?0

Musala Soft Ltd. (2014). *Corporate citizenship and environmental performance report* (pp. 11–13). Sofia, Bulgaria. Retrieved October 12, 2017, from http://www.musala.com/website/wp-content/uploads/2017/01/Environmental_performance_report_2014.pdf

Musala Soft Ltd. (2016). *Corporate citizenship and environmental performance report* (p. 5). Sofia, Bulgaria. Retrieved October 12, 2017, from http://www.musala.com/website/wp-content/uploads/2017/01/Environmental_performance_report_2016.pdf;

Naden, C. (2017). *The rise of being "social"*. Geneva: ISO. Retrieved July 10, 2017, from https://www.iso.org/news/ref2204.html

National Income Agency of Republic of Bulgaria. (2015). *Reference for paid taxes and fees* (p. 1).

Prizma. (2017, April 04). *Foreign investment uncovered*. Retrieved October 10, 2017 form http://investicii.prizma.mk/en/inv/Musala-Soft

Roberts, J. (2010). *How corporate social responsibility (ISO 26000) mandates undermine free markets*. Washington, DC: Heritage Foundation. Retrieved October 13, 2017, from http://www.heritage.org/node/13644/print-display

The European Business Awards. (2017). *Ruban D'Honneur recipients for 2016/17*. Retrieved October 12, 2017, from http://www.businessawardseurope.com/RDH/2016

The OECD. (2003). *Glossary of statistical terms, environmental costs*. OECD. Retrieved from https://stats.oecd.org/glossary/detail.asp?ID=819

The Rio Declaration on Environment and Development. (1992, June 3–14). *The final text of agreements negotiated by governments at the United Nations conference on environment and development (UNCED)*. Rio De Janeiro/New York, NY: Brazil/United Nations Dept. of Public Information, 1993.

Ninel Nesheva-Kiosseva, PhD is a professor of Economics at New Bulgarian University in Sofia, Bulgaria. She is a visiting professor at the University of Bologna-Rimini Campus, University of Urbino and University of Aveiro. Her research interest includes: financial crises, development of banking systems, institutional economics, management accounting, sustainability, social and environmental accounting and reporting, pricing of public utilities, regulation of markets. She is a member of the Bulgarian Macroeconomic Association. She has published in international journals as well as in national and international conference proceedings and books.

Part III
Standardised CSR Facts on ISO 26000

Expectations Versus Applications: Five Years with ISO 26000 in Norway

Caroline D. Ditlev-Simonsen

1 Introduction

In November 2010 the ISO 26000 was launched by the International Organization for Standardization (ISO). The ISO 26000, also called ISO SR, is a guideline for social responsibility. In all, 450 experts from 99 countries and 40 international organizations contributed to develop ISO 26000, and the process took 5 years. Through this period more than 25,000 written comments were addressed, and eventually the participants agreed on the final document (ISO 2016b).

According to ISO, ISO 26000 is "probably the best Social Responsibility standard in the world" (ISO 2016b). When ISO 26000 was launched in Norway in December 2010, the Minister for Trade and Industry claimed that the standard would be a useful guide for both private and public business (Standard Norge 2010a).

This article studies the expectations for ISO 26000 from two leading Norwegian companies in 2010, and compares it to the actual results by 2016.

First, I will give an introduction to ISO 26000. I will then present and compare the development of ISO 26000 application in the two companies from 2011 to 2016. I will conclude with the contributions of ISO 26000 and what other initiatives can learn from the ISO 26000 history of the two companies.

C. D. Ditlev-Simonsen (✉)
BI Norwegian Business School, Oslo, Norway
e-mail: caroline.d.ditlev-simonsen@bi.no

S. O. Idowu et al. (eds.), *ISO 26000 - A Standardized View on Corporate Social Responsibility*, CSR, Sustainability, Ethics & Governance,
https://doi.org/10.1007/978-3-319-92651-3_12

2 ISO 26000

ISO SR is a product of comprehensive cooperation globally over several years. From a positive perspective, this implies that ISO SR is a product of consensus. However, when many stakeholders are involved, the products might get rather “tame” and vaguely formulated (too many compromises). It can be argued that this is the case for ISO 26000, too. On the other hand, developing a common understanding of terms and content for different stakeholders can add value and be crucial for coordinating development in fields like corporate social responsibility.

When ISO 26000 was launched in 2010 it offered unique guidance on how to address social responsibility in a company or an organization. Still, unlike other ISO products, ISO SR did not offer any certification, but is purely voluntary.

ISO SR identifies two fundamental practices of social responsibility: recognizing social responsibility, and stakeholder identification and engagement. Furthermore, ISO SR identifies seven core subjects of social responsibility:

1. organizational governance;
2. human rights;
3. labour practices;
4. the environment;
5. fair operating practices;
6. consumer issues; and
7. community involvement and development

A six-step process on how to integrate social responsibility throughout an organization is also identified (Standard Norge 2010b).

ISO SR provides an important frame for how to approach CSR through defining social responsibility as well as presenting its history.

Until 2010, responsibility focused on corporations. In the term “Corporate Social Responsibility—CSR,” the “C” was associated with corporations. ISO 26000 pinpointed that not only corporations, but other stakeholders as well, such as government, individuals and organizations, have a social responsibility. Therefore, the “C” (i.e., the corporate element) of CSR was removed. The term used in ISO 26000 was thus “SR,” not “CSR.”

The paper version of ISO SR, a booklet, is sold by national ISO offices worldwide. The book is about 100 pages long, and translated into several languages. It’s available for about 1000 NOK (about EUR100).

Since ISO SR was launched in 2010, one systematic review of the initiative was conducted in 2013/2014. The next Systematic Review of ISO 26000 was launched January 15 2017 (International Organization for Standardization 2017), and the ballot will closed June 5.[1] The Post Publication Organisation (PPO) is responsible for this survey, issued by the ISO Central Secretariat to the 160+ ISO member countries. The purpose is to find out if the standard is confirmed, should be

[1]Communication with Project Manager Tina Bohlin, Swedish Standards Institute May 16, 2017.

withdrawn, or needs to be amended. Seventy percent of those who responded were in favor of confirming the standard (ISO 26000 PPO SAG N 34 2014). However, the response rate was not conveyed in the survey.

Based on the 2013/2014 review, a document identifying future areas of possible revisions was developed. These are some key elements of the issues to be considered in the next revision of ISO SR: Make IS SR shorter and easier to read and use, also for SMEs and NGOs; improve guidance on identifying significant and prioritized issues; and improve guidance on risk management, health, international norms of behavior, measuring impact and communicating how to use ISO SR. After the last review, the next systematic review was confirmed to be conducted within 3 years, thus during 2016 (ISO 26000PPO SAG N 41 2015).

The ISO 26000 website is financed by the 2016 Environmental Award from The Swedish Association of Graduate Engineers and Swedbank Robur. One person is dedicated to following up the development of ISO SR at the Swedish Standards Institute (SIS).

Because ISO 26000 is not associated with a certification, it is difficult to measure the number of companies and organizations applying the ISO 26000 guidance. What can be measured is the number of ISO SR booklets sold. Still, I have not been able to find any overview of the number of ISO 26000 guidebooks sold internationally.

ISO 26000 has received varying degrees of attention in different countries, in part because ISO 26000 is driven by the national ISO offices. If the national ISO office has worked actively on marketing ISO 26000, the chance is that ISO 26000 is used more in those countries. Brazil and Sweden are examples of such countries.

The ISO office or representative in Norway is Standard Norway. This company is responsible for handling and selling ISO Guidelines like ISO 26000. According to Standard Norway, a total of 1286 copies of the ISO 26000 guide have been sold in Norway. Most of these copies were sold around the time the standard was launched in 2011, but copies are still sold once in a while. Standard Norway summed it up by claiming that the "sales [of ISO 26000] cannot be said to be overwhelming. . ..".[2] Given that there are more than 500,000 companies in Norway, a sale of slightly more than 1000 copies confirms this claim.

3 Literature Review

As it is less than 6 years after ISO 26000 launched, not many academic studies on ISO 26000 have been conducted. In addition to ISO PPO's survey, aimed at the ISO member countries, a few studies have been generated by ISO. Whereas ISO's own surveys on ISO 26000 reveal general positive feedback on ISO SR, several independent academic studies suggest another story.

[2]«Vi kan se at det omsettes ett eksemplar nå og da, men omsetningen skriver seg i hovedsak fra da standarden var ny. Omsetningen kan vel ikke sies å være overveldende. . .».

According to the study "Standardizing Social Responsibility? New Perspectives on Guidance Documents and Management System Standards for Sustainable Development," ISO SR's main contribution was to provide a common understanding of corporate responsibility, but did not succeed in facilitating management routines and practices leading to social responsibility (Hahn 2012).

In the study "Signaling strategies for ISO 26000: a firm-level approach," Moratis pinpointed two ISO SR challenges. ISO 26000 is noncertifiable, and is a guidance document rather than a management system, which makes ISO 26000 problematic for firms to use (Moratis 2016).

Hemphill's study "The ISO 26000 guidance on social responsibility international standard. What are the business governance implications?" summarizes the contribution of ISO SR this way: "From a general business governance perspective, the ISO 26000 SR international standard is handicapped by it being too broad in scope to be useful in the context of specific industries and sectors, too costly and time-consuming for many small and medium-sized enterprises to implement, and, unlike most other ISO international standards, it is not a certifiable management system—therefore leading to weaknesses in assessing its efficacy" (Hemphill 2013).

4 Data and Method

ISO 26000 was launched in Norway in 2010 by Standard Norway. Through an event hosted by one of Norway's leading certification consultancy companies, the Minister of Trade and a major consulting company claimed that the ISO 26000 was an important and valuable initiative. This triggered interest in studying the choice of signing up for ISO 26000 and the associated expectations.

In 2011 the ISO office in Norway, Standard Norway, was contacted to obtain the names of Norwegian companies that claimed to use or had signed up to use ISO 26000. Two companies were suggested, and these were contacted through the person in charge of CSR and/or sustainability. Both these persons were willing to share their approaches to and expectations for ISO 26000. Separate interviews with both companies were conducted in March 2011 (first interview round).

This study thus applies the case method, and the selection of the two cases is based on the fact that they were the ISO 26000 frontrunners in Norway. Through observations related to ISO 26000 in the two companies' similarities and differences were detected. This inductive case approach is a recognized and useful tool for greater understanding and, eventually, answering the research question in this type of study (Eisenhardt 1989; Yin 2003).

In June 2016 (5 years later), the same companies were contacted and again were willing to share their experience with ISO 26000 (second interview round).

This study follows the two companies from when they started using ISO 26000 in 2011, their experience with ISO 26000 over last 5 years, and the status of ISO 26000 application as of 2016.

5 2011 First Interview

The format for the interview was based on the theory of knowledge transfer as translation (KTT), a concept developed by Røvik (2007). The model is based on pragmatic institutionalism, positioned between the modernistic and the social constructivist paradigm. It focuses on how individuals shape and transform ideas. This individual is named "the translator." The "translator" and person interviewed in this study is the person in charge of ISO 26000 in the two companies, the Head of CSR/Sustainability.

There are four key elements, and thus questions, in the KTT model. Below I have listed these four questions, and what the "translator" interviewed was asked with regard to ISO 26000 (in italics).

1. The translator's motivation: *Why did the "translator" decide to use ISO 26000?*
2. What are the characteristics of the translator: *What are the translator's background, competence and position?*
3. Translation rules: *What was the process of applying ISO 26000?*
4. The effect of the translation: *What has been the effect so far of applying ISO 26000?*

6 2016 Second Interview

The second interview was conducted in June 2016. In one of the companies, the same person as in 2011 was Head of CSR/sustainability. In the second company the person interviewed in 2011 had left the company, so the person who had taken over the position was interviewed in 2016.

The framework for this interview was mainly following up on point four of the KTT interview model. The interview subjects were also asked to come up with suggestions for future development of ISO 26000.

Given that in 2011, ISO 26000 had been launched only a few months earlier, it was hard to measure the effect of the ISO guide at the time. In 2016, more than 5 years since the company started to use ISO 26000, the translators knew much more about the status and effect of ISO 26000. The focus of the 2016 interview was what "happened" to ISO 26000 in the companies and why, and how the translators perceived the future of ISO 26000. The translators were thus asked about the reason for the current status and to come up with recommendations for the future of ISO 26000.

Both employees interviewed in 2016 were explicitly asked *not* to prepare for the interview. The point was to study their current knowledge and use of ISO 26000.

After each interview the interview subject received a summary of the interview to have the opportunity to revise and ensure that the author had not misunderstood.

7 ISO 26000 in Norway: Two Cases

To ensure anonymity, I have named the two companies Alpha and Beta. Below is the result of the interviews in 2011 and 2016.

7.1 Company Alfa

First Interview: In 2011

Alpha was a large Norwegian company in the finance industry and a frontrunner in the CSR field. The "translator" (the person interviewed) was head of CSR in the company, and prior to this position, had been involved in the process of developing ISO 26000. The person was thus well acquainted with ISO 26000. The multi-stakeholder perspective of this product as corporate guidance in the CSR process was especially valuable. The translator perceived the involvement of developing countries in the process as a great advantage.

The translator perceived ISO 26000 as a guidance tool, and therefore did not consider it necessary to get approval from top management and board level to use it. This is different from the processes associated with signing up for UN Global Compact. Such a signature has to be approved at a corporate management level. It was up to the head of CSR to decide whether or not to use ISO 26000. The translator decided to use ISO 26000 as the person's guide and framework for implementing CSR in the company. The company's CSR communication, for example, was following the ISO 26000 structure. The translator also used the ISO 26000 as a guide and handbook for CSR.

Even before ISO 26000 was available, the company had a well-established CSR program. Still, according to the head of CSR, the company worked on continuously improving its CSR program, and ISO 26000 was perceived as a guide and a contribution to this work.

In this company, ISO 26000 was perceived as a contribution to the development of CSR based on the already ongoing CSR program, and as a checklist for this work.

Second Interview: In 2016

Alpha is still a large Norwegian company and a leader in the CSR field. The head of CSR in 2011 recently left the company and a new person took over the position a few months earlier. This new person was interviewed and thus the 2016 "translator." The translator had worked in the CSR field a long time before working for Alpha.

The translator knew little about ISO 26000, and was not aware that the guide was being used in the company, except as a reference for the company's definition of CSR. In ISO 26000, CR is defined as:

social responsibility

responsibility of an **organization** [] for the **impacts** [] of its decisions and activities on society and the **environment** [], through transparent and **ethical behaviour** [] that

- contributes to **sustainable development** [], including health and the welfare of society;
- takes into account the expectations of **stakeholders** []
- is in compliance with applicable law and consistent with **international norms of behaviour** []; and
- is integrated throughout the **organization** [] and practised in its relationships (page 3)

This is the same definition used for CSR in the company today. However, today the main CSR guidance tool for CSR is the UN Global Compact, and the Global Reporting Initiative (GRI) is the frame for CSR reporting. These two guidance tools thus cover the company's needs, with no need for ISO 26000.

Since ISO 26000 was launched, the guidance tool has received very little attention. The translator had not been reminded about ISO 26000 through conferences or media. With little heard about ISO 26000 for years, it has been forgotten.

Contrary to CSR tools like the UN Global Compact and GRI, which can be downloaded without registration from the internet, the ISO 26000 Guidelines have to be purchased. Therefore, ISO has contact information about customers and a unique opportunity to follow up with them. Still, as far as the translator was aware of, ISO had not used this opportunity.

According to the translator, the availability of the UN Global Compact and GRI framework for reporting makes the ISO 26000 unnecessary. Furthermore, since ISO 26000 was launched, there has been a rapid development in the field of CSR. Now it is no longer necessary for a general CSR guide, but rather for more specific indicators to measure the company's CSR performance associated with sectors. Help with identifying key materiality elements and prioritizing work is the type of guidance companies need today. ISO 26000 focuses on process, not measurement, and is therefore to a large extent outdated.

If ISO 26000 were to fulfill a need not already in place, it would be to provide some type of verification, along the lines of other ISO standards. This could be a venue of development for ISO 26000. However, the translator is not certain that companies would be inclined to pay for such a service.

7.2 Company Beta

First Interview: In 2011

Beta represents a large, privately owned Norwegian service chain. All the company's properties and facilities are ISO 14001 certified, so using ISO 26000 was a natural step with regard to CSR. At an ISO 26000 workshop arranged by a leading consultancy and certification company, Beta decided to apply ISO 26000. This workshop was organized before the launch of ISO 26000 in 2010. The decision to use ISO 26000 also occurred a few months before the "translator" (the person interviewed) started as Head of Sustainability in the company.

The person said ISO 26000 provided good guidelines. Specifically, advice on involving stakeholders in defining the company's key CSR engagement was useful. Furthermore, the fact that ISO 26000 pinpointed the importance of concentrating CSR engagement in the company's key business areas was valuable. Prior to the translator being hired, the company's CSR work had not been organized in that way.

The company's owner indicated that the company should be a leader in the CSR field. However, the company lacked a good structure for its CSR work, and decisions so far had been unstructured and not fully integrated. Still, the translator said that CSR was perceived as very important in the company. As a tool for structuring and implementing CSR, the translator perceived ISO 26000 as a good option.

The translator believed that ISO would be a good format for the company's CSR work, and contribute to identify new areas of focus in cooperation with different stakeholders.

Second Interview: In 2016

Beta is still a company known for its CSR and sustainability focus.

When joining the company in 2010, the translator had read the entire ISO 26000 manual carefully and used the seven key core elements as a basis for mapping the sustainability status of the company.

At that time (in 2010) it was unclear what CSR contained for the company, and therefore a tool like ISO 26000 was important for guidance. ISO 26000 contributed to "framing" what CSR was about, and the seven core elements helped decide how to relate to it. The translator applied ISO 26000 as a 360 evaluation tool to form new goals and strategies within the seven areas of the guide.

When the CSR program was up and working, around 2012, ISO 26000 was no longer explicitly applied. To illustrate how obsolete the ISO 26000 had become, the translator claimed that he/she was not able to find ISO 26000 guidelines in the office.

According to the translator, ISO 26000 was excellent as a textbook at the time it was launched. Today, the UN 17 Sustainable Development Goals (SDG) have taken over the role of providing guidelines for the company's CSR work and serving as a checklist for focus areas. Hence, according to the translator, there is no longer a need for ISO 26000 in its current shape and format.

Since ISO 26000 was launched, many other CSR and sustainability tools have been developed. Whereas ISO 26000 has been static, other tools like the UN Global Compact, Global Reporting Initiative (GRI) and Sustainable Development Goals (SDG) have been developed and adapted over time.

ISO 26000 could still provide certification, along the lines of similar ISO products, such as ISO 14001. A major challenge for Beta today is collecting and evaluating information provided by suppliers' sustainability status. As a supplier of services, Beta also has to answer lots of questions and fill out formulas related to CSR and sustainability in order to be accredited for delivery. If ISO 26000 certification could help reduce this paperwork, it would be of great interest. Relative to how much time Beta would save on paperwork due to an ISO 26000 certification, Beta would also be interested in paying for such a certification.

7.3 Analysis

In the analysis I will investigate differences and similarities related to application of ISO 26000 in the two companies, and use this as a "tool" for understanding the development and effect of ISO 26000, as well as future potentials.

In 2011

Common for the two companies is that the Heads of CSR both were positive about ISO 26000. As the ISO 26000 initiative was taken by the Head of CSR in Alpha, it is not surprising that this person was positive about ISO 26000. In Beta, however, the decision to use ISO 26000 was already taken before the translator started in the company. Still, this person was also positive about ISO 26000.

For Alpha the familiarity and background for using ISO 26000 was based on participation in the development of ISO 26000, whereas in Beta, the reason for using ISO 26000 was based on good experience with ISO 14001, and encouragement/invitation by consultancy companies to be one of the frontrunners in applying ISO 26000.

Even though the background for applying ISO 26000 was very different for the two companies, both used ISO 26000 as a framing tool for their CSR work. More specifically, the seven core areas presented in ISO 26000 were used as the guide for the companies to identify what they were to take into account in their CSR work. Both Alpha and Beta used these core areas as a checklist for their CSR work. Alpha also decided to use the ISO 26000 as a framework for its CSR reporting, and the CR definition in ISO 26000 was applied as the company definition for CSR.

In 2016

Common to both companies in 2016 is that ISO 26000 was no longer actively used. Both Alpha and Beta mentioned that ISO 26000 was rarely mentioned among colleagues in the CSR field, nor was it explicitly visible at conferences, media, sector arrangements or similar situations. The fact that none of the Heads of CSR/Sustainability knew where in their office they could find the ISO 26000 Guidelines illustrates the non-use of the guidelines. In Alpha, however, ISO 26000 was still relevant, applied as the ISO 26000 C(S)R definition was still used to guide what CSR implied for the company. This definition is an important guide for what the companies' CSR work entails.

In Beta, ISO 26000 remained to a lesser degree in the company, as UN Sustainability Goals had taken over as the company's checklist for CSR engagement.

Both companies agreed that news about ISO 26000 had been very "quiet," and they were rarely reminded of its existence. Furthermore, both agreed that ISO 26000 had not developed further like other CSR tools such as GRI and UNGC. As there had been a rapid development in the CSR field, tools needed to develop accordingly. In that respect, ISO 26000 was outdated.

For a continuance or revitalization of ISO 26000, both Alpha and Beta suggested developing ISO 26000 into a certification standard. There is a need for such a standard, and as ISO is recognized as a high-quality standard manager, ISO 26000 would have an advantage over other similar initiatives.

8 Discussion and Conclusion

The purpose of this study was to investigate expectations versus applications of ISO 26000 over a period of 5 years; i.e., from when ISO 26000 was launched. Both the companies followed used ISO 26000 as a CSR guidance tool. The focus was on CSR definition and the seven core points as a checklist.

In that sense ISO 26000 has worked as a CSR guide, which was its purpose, and can thus be evaluated as a success. On the other hand, as ISO 26000 five years later is no longer actively applied in the same two companies, can it be perceived as a failure? Most ISO products and other guidance and labeling products are made to be used by an increasing number of companies or individuals. Many such initiatives and labels do not reach this goal, and are gradually forgotten and cease to exist. Many models and theories can be consulted or applied to try to explain why some initiatives survive and flourish, whereas others cease. That was, however, not the purpose of this study, which evaluates the expectations versus the applications.

Whereas the UN Global Compact (UNGC) and Global Reporting Initiative (GRI), the ISO 26000's main "competitors," have grown in popularity and developed over time, ISO 26000 has not. From the ISO 26000 website, we can see how ISO 26000 tries to link up with UNGC and GRI through such publications as "An Introduction to Linkages between UN Global Compact Principles and ISO 26000 Core Subjects" (United Nations Global Compact 2010) and "GRI G4 Guidelines and ISO 26000:2010; A Guide to how to use the GRI G4 Guidelines and ISO 26000 in conjunction," (ISO 2016a) as well as "ISO 26000 and the International Integrated Reporting (IR) Framework, Briefing summary" (ISO 26000 2015). However, for the two companies studied in Norway, this is of less relevance, as ISO 26000 is no longer in use.

The immediate finding of this study is to conclude that in many ways ISO 26000 has been "forgotten" over a period of 5 years in Norway. In addition, the study contributes to knowledge in the field on the effect of voluntary measures and to what extent they flourish or cease, and whether or not they contribute to sustainable development. Further studies should look into whether or not there are different stories of the expectations versus application of ISO 26000 in other countries, and why.

Another interesting study would be to compare the history of ISO 26000 in Norway with the history of another voluntary responsibility and sustainability measure, such as the UN Global Sustainable Development Goals (SDGs). The

SDGs were launched in 2015 when UN Secretary General Ban Ki-moon claimed that "The seventeen Sustainable Development Goals (SDGs) are our shared vision of humanity and a social contract between the world's leaders and the people." Applying the same approach as this ISO 26000 study, the expectations versus application of the SDGs can be investigated. Were their histories the same, or different? And why? These are interesting research questions that can contribute to knowledge about what makes some initiatives work whereas others fail or are "forgotten."

From a practical point of view, this study is relevant for companies to understand corporate approaches to voluntary tools and guidelines and to benchmark themselves. Also, given the history of ISO 26000 provided by this study, companies can become more conscious about why, how and with what expectations they get involved or support different voluntary initiatives.

References

Eisenhardt, K. M. (1989). Building theories from case study research. *Academy of Management Review, 14*(4), 532–550.

Hahn, R. (2012). Standardizing social responsibility? New perspectives on guidance documents and management system standards for sustainable development. *IEEE Transactions on Engineering Management, 59*(4), 717–727.

Hemphill, T. (2013). The ISO 26000 guidance on social responsibility international standard: What are the business governance implications? *Corporate Governance: The International Journal of Effective Board Performance, 13*(3), 305–317.

International Organization for Standardization. (2017). *ISO 26000 – Social responsibility.*

ISO 26000. (2015). *ISO 26000 and the international integrated reporting IR framework briefing – summary* [Press release]

ISO 26000 PPO SAG N 34. (2014). In ISO (Ed.) Advice to ISO/TMB regarding the 2013–2014 systematic review of ISO 26000:2010. Stockholm: ISO. http://iso26000.info/wp-content/uploads/2016/04/N34-PPO-SAG-Recommendations-to-ISOTMB-on-SR-ISO-26000.pdf

ISO 26000PPO SAG N 41. (2015). In ISO (Ed.) PPO protocol: Prioritized areas for revision of ISO 26000:2010. Stockholm: ISO. http://iso26000.info/wp-content/uploads/2016/04/N41-PPO-SAG-PPO-Protocol-Prioritized-areas-for-revision-of-ISO-2600...pdf

ISO. (2016a). *ISO 26000 – Social responsibility.*

ISO. (2016b). In ISO (Ed.) ISO 26000 information. www.iso26000.info

Moratis, L. (2016). Signaling strategies for ISO 26000: A firm-level approach. *International Journal of Operations & Production Management, 36*(5), 512–531.

Røvik, K. A. (2007). *Trender og Translasjoner – Ideer som former det 21. århundrets organisasjon.* Universitetsforlaget. ISBN 978-15-01116-5.

Standard Norge. (2010a). *Ny internasjonal standard for samfunnsansvar: Verdens viktigste standard.* http://www.mynewsdesk.com/no/pressreleases/ny-internasjonal-standard-for-samfunnsansvar-verdens-viktigste-standard-562427

Standard Norge. (2010b). *Veiledning om samfunnsansvar NS-ISO 26000:2010 Guidance on social responsibility: Standard Online AS.*

United Nations Global Compact. (2010). *An introduction to linkages between UN global compact principles and ISO 26000 core subjects.* https://www.unglobalcompact.org/docs/news_events/8.1/UNGC_ISO_Final.pdf

Yin, R. K. (2003). *Case study research: Design and methods.* Thousand Oaks, CA: Sage.

Caroline D. Ditlev-Simonsen, PhD is professor of corporate social responsibility (CSR) at the Department of Law and Governance at BI Norwegian Business School. Her current research focus is profitable CSR, social and environmental impact assessment of corporate activities, socially responsible investment (SRI and ESG), attitudes and behavior associated with sustainable development, leadership as well as teaching methods for sustainable development. For an overview of her academic contributions, please see publications.

Ditlev-Simonsen has international and comprehensive business and organizational experience in the areas of corporate responsibility, corporate citizenship and environmental and ethical issues. Previous work experience includes Project Manager, World Industry Council for the Environment, New York; Executive Officer, Norwegian Pollution Control Authority; Advisor, Kværner ASA and Vice President, Head of Community Contact, Storebrand ASA. She has varied board experience and was a board member of WWF-Norway (World Wide Fund for Nature) from 2002 to 2008.

Her international and comprehensive business and organizational experience, combined with research, contributes to knowledge that is not only published in recognized publications, but also can be of practical use in business strategy and day-to-day business.

Ditlev-Simonsen holds a PhD in Leadership and Organization (BI Norwegian Business School), a Master degree in Energy and Environmental Studies (Boston University) and a Bachelor degree in Business Administration (Simon Fraser University, Vancouver). She is also Co-Director at the BI Centre for Corporate Responsibility, Head of BI Centre for Foundations, member of Appeal Board for Norwegian Foundations (Stiftelsesklagenemda) and member of the Board of Trustees (Styremedlem) of BI Norwegian Business School.

Organisational Challenges to Corporate Social Responsibility

Rafael Mattos Deus, Bruno Michel Roman Pais Seles, Karina Rabelo Ogasawara Vieira, and Rosane Aparecida Gomes Battistelle

1 Introduction

In today's world, globalization has created an interdependence between organizations from varied spheres. In this context, international standards and norms can be helpful during exchanges between countries, guaranteeing the quality, compatibility, and technical feasibility of the products, and this has been formalized into the International Organization for Standardization (ISO), a non-governmental organization created in 1947 (Ward 2011).

Initially, in the 1980s, ISO 9000 was developed, referring to quality management standards, which was widely accepted. In 1996, ISO 14000 was created, which is related to environmental management (Castka and Balzarova 2008). In 2010, ISO launched ISO 26000, an ambitious standard providing guidance on integrating social responsibility into organizations (Ward 2011).

Because of the pressures arising from the globalization of non-governmental organizations (NGOs), corporate social responsibility (CSR) became part of the sustainability debate within organizations (Schwartz and Tilling 2009). CSR aims to generate economic, social, and environmental benefits (i.e., to contribute to the three pillars of sustainable development defined by the United Nations Conference on Environment and Development in 1992) (Elkington 1994; Hart and Milstein 2003).

R. M. Deus (✉) · B. M. R. P. Seles · K. R. O. Vieira
School of Engineering, Department of Production Engineering, São Paulo State University (UNESP), Bauru, Brazil

R. A. G. Battistelle
School of Engineering, Department of Civil and Environmental Engineering, São Paulo State University (UNESP), Bauru, Brazil
e-mail: rosane@feb.unesp.br

S. O. Idowu et al. (eds.), *ISO 26000 - A Standardized View on Corporate Social Responsibility*, CSR, Sustainability, Ethics & Governance,
https://doi.org/10.1007/978-3-319-92651-3_13

While NGOs seek to denounce organizations that exploit people for greater profit (Schwartz and Tilling 2009), Banerjee (2012) warns that companies cannot assume the role of governments to ensure social welfare. ISO 26000 promotes the universality of social responsibility, proposing that both the State and organizations are responsible for social development (International Organization for Standardization 2010), as well as the reduction of poverty and promotion of sustainable livelihoods (Siegele and Ward 2007).

ISO 26000 is a non-certifiable standard, which aims to assist organizations and to contribute to sustainable development, as well as to employ international standards of behavior (International Organization for Standardization 2010). The standard is governed by the principles presented below and has seven core subjects (Table 1).

1. Accountability: the organization must take responsibility for its impacts on society, economy, and the environment;
2. Transparency: acting with transparency in the decisions and organizational activities that impact society and the environment;
3. Ethical Behavior: behave with honesty, fairness, and integrity. Adopt and apply ethical standards of behavior according to the organizational activities developed;
4. Respect for the stakeholder's interests: identify all stakeholders and respect their legitimate rights, and consider other interests of all individuals, not just owners and shareholders;
5. Respect for the rule of law: comply with the laws of all jurisdictions, keeping up-to-date to always comply with the law;
6. Respect for international norms of behavior: in situations where the legislation does not present norms of adequate social and environmental protection, it should at least respect international norms of behavior and avoid being complicit in activities that do not respect international norms of behavior;
7. Respect for human rights: respect and, if possible, promote the rights provided for in the International Declaration of Human Rights. In situations where human rights are not protected, never take advantage of such situations and respect international standards of behavior.

Knowing what social responsibility means is not the same as being aware of practicing social responsibility. Valmohammadi (2011) points out the lack of knowledge and awareness about CSR as a challenge.

Hasan and Almubarak (2016) indicated that while the owners and managers of small- and medium-sized enterprises in Bangladesh have a high level of understanding of the concept of social responsibility, their entrepreneurial actions are for profit. According to the interviewees, this is because they are part of a culture of corruption in which social responsibility faces many challenges.

Salazar et al. (2017) found that people working in a Mexican fumigation company understand social responsibility in accordance with respect for the environment, the quality of working conditions, ethics employed in relationships with stakeholders, the respect for the consumer, and the degree of participation in the community dedicated by the company. Despite the difficulty of generalizing a case

Table 1 The responsibility of the organizations according to the core subject of ISO 26000

Core subject	Organization responsibility
Organizational governance	Have a governance system that practices and oversees the principles of social responsibility. Establishing a culture and environment with hands-on activities that reflect the commitment. Strategies, objectives, and targets should be established under a vision of commitment to social responsibility, balancing the immediate and future needs of the organization and its stakeholders. It is up to the organization to involve all levels of workers by balancing the level of authority, responsibility, and capacity of the decision makers in the company. Analyze and evaluate processes of governance of the organization by adjusting them according to the results and communicating the changes throughout the organization
Human rights	Respect the International Declaration on Human Rights and fundamental labor rights identified by the International Labor Organization (ILO). In a proactive perspective to join efforts with the government to make feasible the fulfillment of economic, social, and cultural rights; facilitate community access to education; and adapt goods and services to the purchasing power of the poor, etc.
Labour practices	Compliance with basic labor standards, principles, and rights established by the International Labor Organization (ILO) contributes to ensuring that companies do not engage in abuse and unfair competition. Companies should also follow the laws established by their countries, regardless of the requirements or support offered by governments. Work in the company must be performed by men and women legally recognized as employed or as self-employed. Employees should be treated equally with no discrimination, receiving the information necessary to perform the routine activities. When the company establishes a code of work practices that must be observed by suppliers and outsourced, it must be consistent with the Universal Declaration of Human Rights and international labor standards
The environment	Manage the environment. Take responsibility for the environmental impacts caused by organizational activities. Obey legislation and implement effective environmental management. Employ operational practices, principles, approaches, and strategies to prevent pollution, optimize the use of natural resources, and minimize waste, emissions, and effluents. Contribute to the mitigation of climate change, preservation of the environment, biodiversity, and restoration of natural habitats
Fair operating practices	Positive results can be obtained by exercising leadership and promoting the adoption of social responsibility. The company must implement anti-corruption policies and practices, training and raising awareness among its employees, representatives, outsourcers, and suppliers to eradicate bribery and report violations of the organization's policies. Prepare them to deal with political activities and conflicts of interest. The company's organizational policies must be transparent. Practices should influence the supply chain to achieve socially responsible goals. Actions such as political gifts that can be perceived as undue influences

(continued)

Table 1 (continued)

Core subject	Organization responsibility
	should be avoided, as well as avoiding taking advantage of social conditions such as poverty. When companies recognize the right to property, they stimulate creation and innovation
Consumer issues	Follow the United Nations Guidelines for Consumer Protection and the International Covenant on Economic, Social and Cultural Rights. Always provide accurate information, employ fair marketing practices, apply fair, transparent and useful contractual processes, promote sustainable consumption, and invest in the design of products and services that provide access to all. When appropriate, care for the most vulnerable and underprivileged. To seek the minimization of risks arising from the use of products and services, using design, manufacturing, distribution, provision of information, support services, withdrawal of the product from the market, and recall. Protect information security and consumer privacy. Provide customer service
Community involvement and development	Companies need to consider themselves part of the community, acting in the prevention and solution of some problems. They should recognize cultural, religious, traditions, histories, and community needs in interacting with them. The expansion and diversification of economic activities and technological development bring contributions to the community, such as job creation, qualification courses, instruction, and health care. The creation of an environment that brings lasting benefits by promoting income and generating wealth by encouraging entrepreneurship and cooperative. Organizations should formulate policies, implementation and evaluation of development programs, and act with the competent authority or humanitarian organizations in situations of crisis, unexpected disasters, catastrophes. The Copenhagen Declaration and the Program of Action called upon the international community to combat poverty, to achieve the goal of productive employment, adequately remunerated and freely chosen, and to promote social integration as a primary development objective. The United Nations Millennium Declaration sets goals that, if achieved, would help to address the significant global development challenges. The United Nations Millennium Declaration emphasizes that, while development should be guided and driven primarily by public policy, the development process depends on the contributions of all organizations. At the local level, community involvement helps to contribute to the achievement of these goals. The Rio Declaration on Environment and Development introduced Agenda 21, which is a process to develop a comprehensive plan of action that can be implemented locally by organizations in each area where human activities impact society and the environment

Source: Based on ISO 26000:2010

study, the variables cited by the respondents are similar to those of the rankings of the best companies to work for (Great Place to Work 2017).

In this context, by improving its corporate identity might also improve financial performance by fostering principles of sustainable development (Wolak-Tuzimek et al. 2017).

Thus, it becomes clear how essential it is for organizations are to direct their efforts towards sustainable development when planning strategies of social interest, as advocated by the School of Environmental Management (Holt 1999).

2 Drivers Towards the CSR

The drivers for standard adoptions, such as ISO 26000 and OHSAS 18001, as well as CSR initiatives adoption, are diverse and depend on organizational characteristics and interactions in the organizational environment. According to Agudo-Valiente et al. (2017), many drivers for CSR engagement are related to ethical theories. These authors propose that CSR is morally correct, which is why organizations must be socially responsible.

Organizations that have formal strategic planning are more proactive when dealing with commercial and non-commercial issues (e.g., CSR). This is because these organizations monitor their environment and directing-actions for diverse perceived questions, including CSR (Kalyar et al. 2013). Strategic planning permits an organization to have a more in-depth view and comprehension of social responsibilities, which enables the implementation of CSR policies and practices (Kalyar et al. 2013).

Within the organizational environment, stakeholders are important for motivating organizations to adopt or follow social standards and to apply CSR initiatives. Considering the strategic planning of an organization, stakeholders are a critical strategic organizational development (Kalyar et al. 2013). Qi et al. (2013) noted that each stakeholder can influence the adoption of specific standards. For example, foreign clients and the community are essential for the adoption of ISO 9001, and foreign investors are essential for ISO 14001 adoption.

In addition to stakeholders, many other factors influence the adoption of social standards and CSR initiatives (Kalyar et al. 2013). Below, we present the primary drivers for adoption of social standards, such as ISO 26000, and CSR initiatives.

Because of globalization, competition in the international market can be a driver for adoption of ISO 26000 (Castka and Balzarova 2008; Høivik 2011; Valmohammadi 2011, 2014). Multinational organizations adopt ISO 26000 to achieve legitimacy for their intern policies of social responsibility because it facilitates access to different international markets (Castka and Balzarova 2008; Høivik 2011). In this context, the organizations can also find opportunities to be part of international joint-ventures, which can be considered as a driver for adopting ISO 26000 (Valmohammadi 2011, 2014).

Reputation is a significant driver for the adoption of ISO 26000 because the clients and others stakeholders pay attention to proactive actions of social responsibility (Pojasek 2011). Therefore, these proactive actions can improve the relationship between organizations and clients (Høivik 2011) and, consequently, improve the organization reputation (Valmohammadi 2011). To illustrate, Hasan and Almubarak (2016) studied small-medium enterprises (SMEs) engaged in socially responsible business practices in Bangladesh and discovered that 18% of their research sample considered “to improve their overall business reputation” as a driver for the engagement in social responsibility.

As explained earlier, the stakeholders are essential for motivating an organization to adopt or follow social standards and to apply CSR initiatives. Some studies have proposed that adoption of ISO 26000 could improve the relationship with external stakeholders. In specific cases, the stakeholders might notice that some organizations need to adopt social standards or CSR initiatives (Castka and Balzarova 2008); consequently, the stakeholders might pressure organizations to adopt more socially-friendly practices and actions. It is stakeholder’s role to exert pressure on organizations and managers to adopt more socially-friendly actions, as well as to disclose extra social information (Habbash 2016). Moreover, ISO 26000 presents solutions to solve conflicts between organizations and stakeholders, as well as opportunities to reinforce their relationship (Høivik 2011; Høivik and Shankar 2011). The standards also improve the organization’s capacity to maintain clients and to develop a responsible behavior with them (Merlin et al. 2012; Valmohammadi 2011).

Competitive advantage is perhaps a more important motivator for the adoption of social standards. Some studies present a variety of competitive advantages from social certification adoption. For example: advantage when confronted with competitors, increased innovation (the development of new business), increased social innovation [as pointed out by Harazin and Kósi (2013)], new business opportunities, and an increased ability to achieve desired financial results (goals) (Castka and Balzarova 2008; Hahn 2012; Harazin and Kósi 2013; Høivik 2011; Høivik and Shankar 2011; Merlin et al. 2012; Ortová and Stanková 2011; Valmohammadi 2011).

Another significant driver is the decrease in risks around the business. Directly related to competitive advantage, this driver is essential because the adoption of CSR initiatives helps to manage risks in business by preventing and mitigating them (Høivik 2011; Valmohammadi 2011).

Adherence to government laws can be a motivator for engaging in social responsibility (Hasan and Almubarak 2016; Valmohammadi 2014). For example, in a study carried out by Hasan and Almubarak (2016), SMEs in Bangladesh considered that “to comply with domestic and international laws” as a driver to engage CSR initiatives.

The drivers presented above were related to external aspects of an organization, such as the interactions in the organizational environment and competitiveness. Below, we present some drivers related to the internal aspects of an organization.

The social standards adoption and CSR initiatives improve the work environment and make it more sustainable (Ortová and Stanková 2011) by improving the relationship between employees and the organization, as well as between co-workers. Moreover, social standards adoption and CSR initiatives increase the organization's chances of attracting talented and desired professionals (Høivik 2011; Valmohammadi 2011).

Organizations that have adopted other international management systems standards could be motivated to adopt ISO 26000 (Castka and Balzarova 2008; Høivik 2011) because social standards are often harmonious with other standards, such as ISO 14001 (Høivik 2011). The integration of management systems can reduce costs, such as deployment costs, and can improve "operational efficiency, employee motivation, efficient management and utilization of organizational resources, gaining competitive advantage and for sustainable development" (Muthu et al. 2015).

Company size can be a driver to adopt or follow the content of ISO 26000. Barnes and Croker (2013) identified that social responsibility issues in ISO 26000, especially the health and safety in the workplace and to the environment, are very relevant to Hong Kong construction companies with more than 200 employees.

Finally, ISO 26000 is a guide to the implementation of CSR aspects in diverse types of organizations. This standard offers a guide to improve organization CSR credibility (Pojasek 2011) and could be necessary for some organizations to identify consensus definitions of central themes related to CSR (Hahn 2012).

3 Implementation Barriers

Barriers block and prevent the satisfaction of a need or the locomotion for a specific vector (Chiavenato 2004). Barriers are also encountered when implementing more sustainable organizational practices that move an organization towards a more significant social responsibility. Such blockages might prevent the implementation of organizational practices that would lead to success, growth and development, and improved organizational competition.

It is essential that companies know both what motivates them in a particular direction and what they can prevent. For example, in the face of increasing public environmental awareness, the power generation sector will have to deal with barriers to seeking technological alternatives, even if they are highly risky and with an uncertain future. Other sectors that face barriers to sustainability can also overcome them by developing the capacity for absorption and communicative skills (Pinkse and Dommisse 2009).

It is worth mentioning that specific factors can be barriers for some organizations, while for others they are challenges and a starting point for development (Pinkse and Dommisse 2009).

As the standards (voluntary guidance standard) or guidelines for social responsibility are new and are part of a sustainability dimension, there is an implicit need to understand the main barriers that prevent the integration of sustainability strategies, since generic assessments of why environmental policies end up failing are neither sufficient nor relevant statements (Adams and Ghaly 2007). Thus, it is important to determine and identify the barriers that block the implementation of ISO 26000, thereby preventing locomotion for a particular vector (in this case, a company with more social responsibility).

It is worth noting that other standards also have implementation barriers. Zeng et al. (2007) highlight the following implementation barriers in a study addressing the ISO 9000 family of quality management system standards: (1) short-sighted objectives; (2) excess expectations of the standard; (3) mandatory requirements without compromise; and (4) follow the certification trend.

Abdullah et al. (2013) list five categories of implementation barriers (organizational, resource, cultural, environmental, and technical) for ISO 9000 in a government enterprise in Malaysia.

The lack of culture management and people with difficulties understanding are barriers to social responsibility and achieving a more innovative environment. Matinaro and Yang (2017) emphasize that managers, in the case of the construction industry, cannot manage or create a culture of innovation. Thus, there are significant sector-level gaps in innovation that have a negative impact on social change for sustainability. The same can be applied to a culture of social responsibility; the top-down relationship is typically important for this development. However, Asif et al. (2013) emphasize that to achieve an integrated socio-technical system, the social aspects of companies must be aligned with their technical structures. The development of employees and the continuous improvement for learning and innovation gradually develop the institutional knowledge of corporate social responsibility through both top-down and bottom-up approaches.

The ISO 14000 series of standards related to the environmental management system also present barriers to its implementation. In a study, Biondi et al. (2000) report that, for small and medium-sized enterprises, the primary barrier is economic, followed by knowledge management and lack of specialized human resources. Because of an organization's strategic position, it is essential to identify implementation barriers, regardless of practices, whether of quality with ISO 9000, environmental with ISO 14000 or social with ISO 26000. It is essential to make clear that there are several implementations of a corporate social responsibility system, either by SA8000 or ISO 26000 standards (Chiarini and Vagnoni 2017). In this way, many barriers will be familiar since they are related to social responsibility practices, whereas others are more specific since these standards differ from each other.

To clarify the difference between the standards of social responsibility, Chiarini and Vagnoni (2017) carried out a quantitative study to highlight the differences between the two social responsibility standards. They conclude that ISO 26000 can have a more significant effect on the effectiveness of a CSR system from a strategic point-of-view and SA8000 is based on strategies to comply with laws and resolutions. When the issue involved economic and financial aspects of the organization, neither the SA8000 nor ISO 26000 had a significant effect, probably because the standards focus more on the principles of effectiveness than on efficiency. Nevertheless, according to the results of Chiarini and Vagnoni (2017), SA8000 focuses more on specific stakeholders, such as workers, unions, and nongovernmental organizations (NGOs), while ISO 26000 involves all possible stakeholders. Another point is that SA8000 is deficient in taking into account the market issues and the category of customers, while ISO 26000 adequately addresses customer needs management.

Valmohammadi (2011) surveyed 130 manufacturing organizations in Iran and ranked the main barriers to implementing CSR practices and initiatives: (1) lack of knowledge or awareness of CSR; (2) focus on removing obstacles and achieving short-term goals rather than developing long-term strategies; (3) lack of knowledge about how to implement CSR with the organization's strategy; (4) lack of adequate communication and promotion of stakeholders; (5) private companies believe that the government should be responsible for sustainable development and leadership in this field; (6) acts of charity are always understood to remain private; and (7) xenophobic mentality and in the form of conspiracy.

For the author, 85% of these organizations expect to have a more successful position in the market by improving the corporate identity resulting from the implementation of CSR activities and programs.

Lack of knowledge and awareness is a prominent barrier, which can be overcome through public policies involving government and higher education institutions. It is worth mentioning that most of the seven main areas of ISO 26000 deal with essential aspects to organizations: the environment, human rights, labor practices, organizational governance, fair operating practices, consumer problems, and contribution to the community and society (Park and Kim 2011). Using the ISO 26000 core can enable success by reducing costs and increasing revenue (Valmohammadi 2011).

The impact of ISO 26000 on trade and its relationship with the World Trade Organization can be considered a barrier if this standard has the same effect as the 9000 and 14000 ISO series. Thus, there would be an adverse effect on trade in countries that do not comply with the human rights aspects, as is the case in many exporting countries.

There are fears that ISO 26000 could cause more global pressure on small producers if they do not meet the guidelines of the standard by limiting market access for products from developing countries (Ward 2011).

Høivik (2011) emphasizes two implementation barriers, both related to knowledge management: the ISO 26000 as a process standard is well-suited to the characteristics of the company and therefore becomes company specific; and obstacles related to how to share this knowledge.

In a systematic review, Deus et al. (2014) highlight ten important implementation barriers for ISO 26000: (1) A lack of alignment of the CSR with the organizational strategy: CRS is essential to the company's success, and there is research linking RSC to competitive success. However, efforts are needed in understanding conceptual and theoretical advances and empirical tests between CSR and company performance (Galbreath 2009); (2) Commercial (national and international): because it can limit a companies' access to the market, especially those that do not value human rights; (3) A lack of understanding of ISO 26000: organizations may not understand it as a whole, thinking that it is irrelevant and inapplicable; (4) A lack of communication: related to the lack of ability to effectively communicate and promote ideas with stakeholders (internal and external); (5) A lack of tools: most of the social responsibility implementation tools are designed for multinationals and not for small and medium-sized enterprises, or start-up companies; (6) A lack of sensitivity to the theme: probably because of a lack of understanding of the theme; (7) A focus on the short term: this means short-sighted objectives, because companies want immediate results, while sustainability-related issues take time; (8) Knowledge management: social responsibility is not an easy question to interpret and apply. Moreover, there is a requirement for specific knowledge and management of this knowledge to be transmitted to the organization, making it shareable; (9) Fear of not complying with the standard: because it involves issues such as organizational governance, human rights, labor practices, the environment, fair operating practices, consumer issues and local community involvement/development. This can generate some discomfort, especially for companies that are not aligned with these topics; (10) Financial resources: not being sure of the financial return due to the resources and time spent in the process.

To overcome these barriers is a challenge for organizational management, which must use the necessary tools presented by the standard, as well as fostering support via the drivers. Thus, the organization must plan (through a clear objective), do (through training and knowledge management, executing and collecting the necessary data), check if the goals have been reached and if the barriers have been transposed, and act, while always seeking the continuous improvement, correcting errors, and preventing and detecting new barriers.

4 Conclusion

ISO 26000 is primarily a guide to CSR, providing tools to improve organizational credibility about CSR. This standard allows the identification of consensus definitions of central subjects, issues of social responsibility, and corporate sustainability, which are exciting and relevant aspects of organizations.

We identified much opportunity to study the capacity of organizations to implement ISO 26000. Because of the specific characteristics of each organization, the implementation of ISO 26000 by one organization cannot be copied by another. Each company must tread its own path while learning from the shared experiences of others.

Further research is now needed to investigate how partnerships between universities and industries can contribute to promoting the adoption of social responsibility by organizations. Also, the role of the States and the World Trade Organization in the diffusion of ISO 26000 should be further investigated. We propose that the creation of an ISO 26000 implementation model for companies, emphasizing small and medium-sized enterprises, is also fundamental for new research.

References

Abdullah, S., Razak, A. A., Hanafi, M. H., & Jaafar, M. (2013). Implementation barriers of ISO 9000 within the Malaysian local government. *International Journal of Quality & Reliability Management, 30*(8), 853–876.

Adams, M. A., & Ghaly, A. E. (2007). Determining barriers to sustainability within the Costa Rican coffee industry. *Sustainable Development, 15*(4), 229–241.

Agudo-Valiente, J. M., Garcés-Ayerbe, C., & Salvador-Figueras, M. (2017). Corporate social responsibility drivers and barriers according to managers' perception; Evidence from Spanish firms. *Sustainability, 9*(10), 1821.

Asif, M., Searcy, C., Zutshi, A., & Fisscher, O. A. (2013). An integrated management systems approach to corporate social responsibility. *Journal of Cleaner Production, 56*, 7–17.

Banerjee, S. B. (2012). Corporate social responsibility: The good, the bad and the ugly. *Critical Sociology, 34*(1), 51–79.

Barnes, L. R., & Croker, N. (2013). The relevance of the ISO 26000 social responsibility issues to the Hong Kong construction industry. *Australasian Journal of Construction Economics and Building, 13*(3), 37–50.

Biondi, V., Frey, M., & Iraldo, F. (2000). Environmental management systems and SMEs. *Greener Management International*, (January), 55–69.

Castka, P., & Balzarova, M. A. (2008). ISO 26000 and supply chains—On the diffusion of the social responsibility standard. *International Journal of Production Economics, 111*(2), 274–286.

Chiarini, A., & Vagnoni, E. (2017). Differences in implementing corporate social responsibility through SA8000 and ISO 26000 standards: Research from European manufacturing. *Journal of Manufacturing Technology Management, 28*(4), 438–457.

Chiavenato, I. (2004). *Introdução À Teoria Geral Da Administração: Edição Compacta* (3rd ed.). Rio de Janeiro: Elsevier.

Deus, R. M., Seles, B. M. R. P., & Vieira, K. R. O. (2014). As Organizações e a ISO 26000: Revisão dos conceitos, dos motivadores e das barreiras de implementação. *Gestão & Produção, 21*(4), 793–809.

Elkington, J. (1994). Towards the suitable corporation: Win-win-win business strategies for sustainable development. *California Management Review*, 90–100.

Galbreath, J. (2009). Building corporate social responsibility into strategy. *European Business Review, 21*(2), 109–127. https://doi.org/10.1108/09555340910940123.

Great Place to Work. (2017). *How you're evaluated.* Retrieved from http://www.greatplacetowork.net/best-companies/about-applying-to-best-companies-lists/how-youre-evaluated

Habbash, M. (2016). Corporate governance and corporate social responsibility disclosure: Evidence from Saudi Arabia. *Social Responsibility Journal, 12*(4), 740–754.

Hahn, R. (2012). ISO 26000 and the standardization of strategic management processes for sustainability and corporate social responsibility. *Business Strategy and the Environment, 22*(7), 442–455.

Harazin, P., & Kósi, K. (2013). Social challenges: Social innovation through social responsibility. *Periodica Polytechnica, Social and Management Sciences, 21*(1), 27–38.

Hart, S. L., & Milstein, M. B. (2003). Creating sustainable value. *Academy of Management Executive, 17*(2), 56–67.

Hasan, F. S. M., & Almubarak, M. M. S. (2016). Factors influencing women entrepreneurs' performance in SMEs. *World Journal of Entrepreneurship, Management and Sustainable Development, 12*(2), 82–101.

Holt, K. (1999). Management and organization through 100 Years. *Technovation, 19*(3), 135–140.

International Organization for Standardization. (2010). *ISO 26000: Guidance on social responsibility.*

Kalyar, M. N., Rafi, N., & Kalyar, A. N. (2013). Factors affecting corporate social responsibility: An empirical study. *Systems Research and Behavioral Science, 30*(4), 495–505.

Matinaro, V., & Yang, L. (2017). Towards increased innovativeness and sustainability through organizational culture: A case study of a Finnish construction business. *Journal of Cleaner Production, 142*(October 2016), 3184–3193.
Merlin, F. K., Pereira, V. L. D. V., & Júnior, W. P. (2012). Sustainable development induction in organizations: A convergence analysis of ISO standards management tools' parameters. *Work, 41*(Suppl 1), 2736–2743.
Muthu, S., Gopalakrishnan, C. P. S., & Ammasaiappan, M. (2015). Integrated management systems for better environmental performance and sustainable development – a review. *Environmental Engineering and Management Journal, 14*(5), 985–1000.
Ortová, M., & Stanková, E. (2011). The preparedness of certain companies to implement the ISO 26000 standard. In *International Conference Liberec Economic Forum* (Vol. 10, pp. 372–381). Liberec: TechnickÃ univerzita v Liberci.
Park, J. Y., & Kim, S. W. (2011). Global corporate social responsibility standard, ISO 26000 and its effect on the society. *Asian Journal on Quality, 12*(3), 315–322.
Pinkse, J., & Dommisse, M. (2009). Overcoming barriers to sustainability: An explanation of residential builders' reluctance to adopt clean technologies. *Business Strategy and the Environment, 18*(8), 515–527.
Pojasek, R. B. (2011). ISO 26000 guidance on social responsibility. *Environmental Quality Management, 20*(3), 85–93.
Qi, G., Zeng, S., Yin, H., & Lin, H. (2013). ISO and OHSAS certifications: How stakeholders affect corporate decisions on sustainability. *Management Decision, 51*(10), 1983–2005.
Salazar, L., Alejandra, J. F. O. H., & Manríquez, M. R. (2017). La Responsabilidad Social Empresarial Desde La Percepción Del Capital Humano. Estudio de Un Caso. *Revista de Contabilidad, 20*(1), 36–46.
Schwartz, B., & Tilling, K. (2009). 'ISO-lating' corporate social responsibility in the organizational context: A dissenting interpretation of ISO 26000. *Corporate Social Responsibility and Environmental Management, 16*(5), 289–299.
Siegele, L., & Ward, H. (2007). Corporate social responsibility: A step towards stronger involvement of business in MEA implementation? *Review of European Community and International Environmental Law, 16*(2), 135–144.
Valmohammadi, C. (2011). Investigating corporate social responsibility practices in Iranian organizations: An ISO 26000 perspective. *Business Strategy Series, 12*(5), 257–263.
Valmohammadi, C. (2014). Impact of corporate social responsibility practices on organizational performance: An ISO 26000 perspective. *Social Responsibility Journal, 10*(3), 455–479.
Ward, H. (2011). The ISO 26000 international guidance standard on social responsibility: Implications for public policy and transnational democracy. *Theoretical Inquiries in Law, 12*(2), 665–718.
von Weltzien Høivik, H. (2011). Embedding CSR as a learning and knowledge creating process: The case for SMEs in Norway. *Journal of Management Development, 30*(10), 1067–1084.
von Weltzien Høivik, H., & Shankar, D. (2011). How can SMEs in a cluster respond to global demands for corporate responsibility? *Journal of Business Ethics, 101*(2), 175–195.
Wolak-Tuzimek, A., Tarnawska, J., & Chmiel, M. (2017). The natural environment as an area of corporate social responsibility. *Ochrona Srodowiska I Zasobów Naturalnych, 28*(3), 45–52.
Zeng, S. X., Tian, P., & Tam, C. M. (2007). Overcoming barriers to sustainable implementation of the ISO 9001 system. *Managerial Auditing Journal, 22*(3), 244–254.

Rafael Mattos Deus is a researcher at the Department of Production Engineering, São Paulo State University (UNESP), Brazil. He holds a Bachelor's degree in Business Administration from the UNISEB University Centre and Biology from the São Paulo State University and has a Master's degree in Production Engineering: Operations and Systems Management from the São Paulo State University. His research interests include business administration, environmental management, corporate social responsibility, solid waste management and circular economy.

Bruno Michel Roman Pais Seles is a researcher at the Department of Production Engineering, São Paulo State University (UNESP), Brazil. He holds a Bachelor's degree in Business Administration from the Sao Paulo State University and has a Master's degree in Production Engineering: Operations and Systems Management from the São Paulo State University. His research interests include business administration, operations management, environmental management, corporate social responsibility and sustainable supply chain management.

Karina Rabelo Ogasawara Vieira is a researcher at the Department of Production Engineering, São Paulo State University (UNESP), Brazil. She holds a bachelor degree in Business Administration from the Sao Paulo State University and has a master's degree in Production Engineering: Operations and Systems Management from the São Paulo State University. Her research interests include business administration, environmental management, corporate social responsibility and sustainability in higher education.

Rosane Aparecida Gomes Battistelle is a researcher and full professor at the Department of Civil and Environmental Engineering, São Paulo State University (UNESP), Brazil. She holds a bachelor degree in Civil Engineering; she has a master's degree in Civil Engineering from the University of São Paulo, and she has doctor's degree in Environmental Engineering Sciences from the University of São Paulo. Her research interests include civil construction, alternative materials and in particle board composed of agro-industrial waste, environmental management, corporate social responsibility, solid waste management and sustainability in higher education.

Index

S. O. Idowu et al. (eds.), *ISO 26000 - A Standardized View on Corporate Social Responsibility*, CSR, Sustainability, Ethics & Governance,
https://doi.org/10.1007/978-3-319-92651-3